History of
Electron Microscopy
in Switzerland

Edited by
John R. Günter

Birkhäuser Verlag
Basel · Boston · Berlin

The publication of this book was kindly supported
by the Swiss Academy of Sciences

Library of Congress Cataloging-in-Publication Data

History of electron microscopy in Switzerland / edited by John R. Günter.
p. cm.
Includes bibliographical references.

1. Electron microscopy – Switzerland – History. I. Günter, John R.
QH212. E4H57 1990
578' .45' 09494 – dc20

Deutsche Bibliothek Cataloging-in-Publication Data

History of electron microscopy in Switzerland / ed. by John R. Günter.
Basel; Boston; Berlin: Birkhäuser, 1990

NE: Günter, John R. (Hrsg.)

© 1990 Birkhäuser Verlag Basel
Softcover reprint of the hardcover 1st edition 1990

ISBN-13: 978-3-0348-7205-8   e-ISBN-13: 978-3-0348-7203-4
DOI: 10.1007/978-3-0348-7203-4

CONTENTS

PREFACE

In 1989, the Swiss Society for Optics and Electron Microscopy (Schweizerische Gesellschaft für Optik und Elektronenmikroskopie - Société Suisse d'Optique et de Microscopie Electronique), formerly founded as "Schweizerisches Komitee für Optik - Comité Suisse d'Optique" could celebrate its 40th anniversary.

Already during and mainly just after World War II the then newly invented electron microscopy was introduced also in Switzerland and its importance quickly increased. In 1955, our Society was split into two sections, i.e. for Optics and for Electron Microscopy, both with their own secretaries.

Other foreign Societies for electron microscopy in Europe and all over the world have celebrated their anniversaries in the last few years and held reviews at these occasions.

In view of this and facing the fact that many of the pioneers and founders of our Society might help to record the history of electron microscopy in our country, the board of SGOEM-SSOME has decided to have a short review of its history written and published. This short review has now developed into a book.

I would like to thank here all the authors, who have contributed to this volume very much. My special thanks go to Prof. Dr.John R. Günter, without whose circumspective and energetic work this review of the history of electron microscopy in Switzerland would never have appeared.

It remains the hope that also in the future Switzerland will contribute to the development of electron microscopy and that we will be able to compete on an international level on the forefront of this scientific method.

Richard Guggenheim,
President, Swiss Society
for Optics and Electron
Microscopy

ACKNOWLEDGEMENTS

It is not self-evident that renowned scientists devote their time to writing historical reviews of a scientific method. Therefore special thanks are due to the authors of the following chapters, who have done so in a very competent manner and the collaboration with whom has been a real pleasure. Their individual contributions have been modified only where necessary, in order to reflect the personal views and the styles of the authors as truly as possible. Thanks are also due to the many colleagues who have contributed to this volume by informing the editor about the existence of old documents, records of their companies or universities and personal reminescences. Such contributions are due to (in alphabetical order):

E.B.Bas, S.Böhler (Balzers Union AG), M.Brönnimann, R.Giovanoli, R.Gleyvod (E.Haefely AG), H.Gross, R.Guggenheim, A.Meyer (Diatome AG), R.Signer, R.Wessicken. The editor also wishes to thank the Swiss Academy of Sciences and the Swiss Society for Optics and Electron Microscopy for financial support of this publication, as well as the staff of the publishers for their helpful cooperation.

John R. Günter, editor.

I. INSTRUMENTATION

History of Electron Microscopy
in Switzerland
Edited by John R. Günter
© 1990 Birkhäuser Verlag Basel

THE INSTRUMENTAL CONTRIBUTION OF SWITZERLAND TO THE DEVELOPMENT OF ELECTRON MICROSCOPY A HISTORICAL REVIEW

Werner Villiger
Biozentrum, University
Basel

The Transmission Electron Microscope

The development of the Swiss transmission electron microscope had started during the second world war around 1941/2.

A well known company for measuring technique, the "Trüb, Täuber & Cie. AG" in Zürich (TTC) had been manufacturing and improving the Dufour-oscillograph, but also developing television scanning and electron diffraction tubes, and thus achieved a sound knowledge in the construction of electron devices in metal (high voltage and vacuum systems). The problems connected to vacuum and measuring techniques as well as to electron optics were intensively studied and solved in an expedient way. This also formed an essential precondition for the development of a Swiss electron microscope. Under the direction of Giovanni Induni, a first instrument called test electron microscope was constructed in 1941/2 and used to gain first experiences. This electron microscope (Fig. 1) was equipped with a cold cathode and three electromagnetic lenses (condensor, objective and projective); see also chapter by K.Mühlethaler. Some of the resulting micrographs of fragments of diatoms were presented to the public in May 1944 (1). A more detailed description of the instrument followed about one year later (2, 3). Already in 1945 G.Induni has published the ray path of a successor model with a "mixed" lens system. It included a

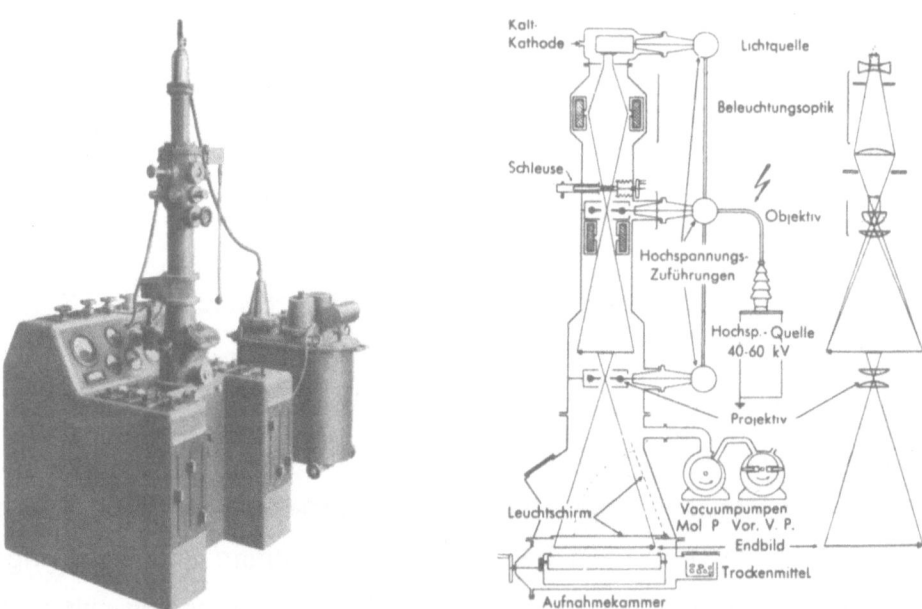

Fig. 1: TTC test electron microscope constructed by G.Induni, 1944-46.
Fig. 2: Comparison of ray paths for the electron microscope and a light microscope. (Both figures published by G.Induni, 1945 (3).

cold cathode with an electromagnetic condensor lens, an electrostatic objective lens, an electromagnetic "focussing lens" and an electrostatic projective lens (Fig. 2). Cathode and electrostatic lenses were operated with a high voltage of 40-60 kV.

The considerably improved resolving power compared to the light microscope which could be achieved with this instrument was a real challenge for biologists, physicians, physicists, chemists, etc. for a deeper penetration into the microcosm.

With the goal to apply and develop the electron microscope and preparation techniques various collaborations between TTC and the Federal Institute of Technology in Zürich (A.Frey-Wyssling, K. Mühlethaler) and the Universities of Geneva (J.Weigle, P.Dinichert) and Berne (W.Feitknecht, H.Studer) have started in 1942-44 (for details see chapters by K.Mühlethaler; E.Kellenberger; J.R.Günter).

Based on experiences with the test electron microscope (only available in the factory) several technical modifications have been realized and improved versions under the designation KM (for "Kaltkathodenstrahl-Mikroskop") were delivered in 1946/47 to the Universities of Geneva (J.Weigle, P.Dinichert and E.Kellenberger) and Berne (W.Feitknecht, H.Studer) and put into service. The Federal Institute of Technology in Zürich (ETH) (A.Frey-Wyssling, K.Mühlethaler) obtained an instrument of a similar type in 1948 (Fig. 3).

The maximum electron optical magnification which these microscopes attained was about 10´000 x. The demands put on these instruments in their practical microscopical use differed greatly from the ideas and experiences of the TTC company, due to the different objects and specimen preparation methods, and led to various technical problems.

6

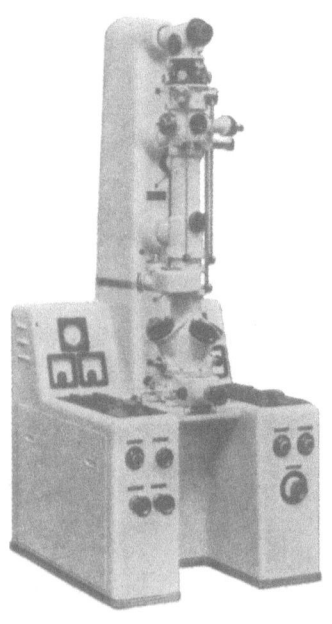

Fig. 3:   TTC electron microscope of the KM type, ETH Zürich, 1948.
Fig. 4:   Modified instrument, University of Geneva, 1952.

To meet these problems, Induni decided in 1946/7 to continue a joint development project with J.Weigle and E.Kellenberger in Geneva, which brought considerable improvements. So, e.g., a new electrostatic lens with an epoxy resin (Cibanit) insulator instead of porcelain was developed in cooperation with the SIP (Société Genèvoise d'Instruments de Physique Genève), showing improved safety of operation as well as a better resolution at the voltage of 50 kV. Following a suggestion of E.Kellenberger and J.Weigle a focussing aid was constructed and tested as well (4). A further modified instrument was installed in 1952 in the same institute in Geneva (Fig. 4).

The experiences and knowledge acquired with the delivered instruments led to the presentation of a newly developed series of microscopes named KM4 by the later scientific director of the TTC company, L.Wegmann (5). One of these microscopes was delivered to the University of Lausanne (A.Gautier) in 1955.

The concept of these instruments was based on the original idea of G.Induni to construct an electron microscope simple and easy to operate and with a resolving power sufficient for routine investigations.

These instruments were mainly used in foreign industries for quality control, distribution analyses, evaluation of materials, e.g. in textile industry etc., and performed well in the investigation of metal shadowed specimens and oxide or Pt/C replicas.

In the investigations of ultrathin sections of biological samples, however, the KM4 yielded unfortunately unsatisfactory results. A direct comparison with electron microscopes of foreign manufacturers equipped only with electromagnetic lenses showed that even with perfect focussing, the TTC-micrographs were not quite sharp.

The main reason for this effect was the energy loss of the electrons in traversing the thin sections. Therefore the University of Basel (G.Wolf-Heidegger, W.Kuhn) ordered an instrument with a high voltage of 70 kV in 1956, and this was put into operation in the same year. First pictures of thin sections taken with this instrument indeed showed an improved image quality and an improved "biological resolution", but as a consequence of the higher voltage, the safety of operation of the instrument was considerably decreased. The electrostatic lenses (which were constructed for 50 kV) as well as their high voltage feedthroughs were repeatedly damaged by discharges and high voltage breakdowns and had to be permanently revised. To meet this high liability to breakdowns, W.Villiger in 1957 tentatively installed a watercooled oil diffusion pump instead of the standard molecular pump. This and a number of other predominantly vacuum technical measures were very efficient and formed the basis for a new joint development of the electron microscope of the type KM5 by W.Villiger, M.Thürkauf and the Trüb, Täuber & Cie. AG (Fig. 5).

This development included a new electrostatic lens for a high voltage of 70 kV, which exhibited much reduced flashover and discharge and consequently a high operational safety and stability due to the use of the new insulating material Haefelit PQ (epoxy resin mixed with quartz sand).

The accumulators (12 V/80 A) used so far for the feed of the electromagnetic lenses of the TTC electron microscopes were replaced by a transistorized mains supply for stabilizing the lens current. This development was made in cooperation with the Department of Physics of the University of Basel.

This instrument was presented for the first time at the Congress on Electron Microscopy in Berlin in 1958 and described in detail later (6).

The KM5 microscope installed in Basel could, after its reconstruction, routinely be used for a number of different investigations. Although for work with ultra thin sections of biological material embedded in plastics the results were improved, they were still not fully satisfactory.

The reasons for this were the following:

1. The electrostatic lenses had an excessive chromatic aberration.

2. The astigmatism could not be corrected because of the lack of a stigmator.

3. The local electron emission of the cold cathode was unstable, becoming apparent at magnifications of about 35´000 x.

Fig. 5:   Prototype of the KM5 electron microscope, University of Basel, 1956/7.

The members of the laboratory for electron microscopy of the University of Basel (W.Villiger/M.Thürkauf) therefore suggested to the TTC company in 1960 to develop jointly a routine microscope built with electro-magnetic lenses only and with a thermal cathode. The idea was enthusiastically accepted and supported by L.Wegmann of TTC. However, the realization of this project failed, because the management of TTC insisted for prestige reasons on the use of the cold cathode.

As a consequence of this position, the University of Basel ceased further cooperation in development with TTC and installed an instrument with electromagnetic lenses and thermal cathode, the Elmiskop I of the Siemens company in Berlin, in 1961.

The production of transmission electron microscopes of the Trüb, Täuber & Cie. AG was definitively stopped at the end of 1960.

Relatively little is known about the later fate of the TTC microscopes. Some of them have been disassembled and parts of them, e.g. lenses etc. are used as demonstration material in teaching.

Of the two instruments delivered to Geneva, it is known that they were rebuilt by E.Kellenberger, J.Bron and D.Karamata into a single instrument in 1960, equipped with a diffusion pump. This instrument under the name "Babar" can still be seen in the Musée d´art et d´histoire in Geneva (Fig. 6).

Another instrument, the KM5 of the University of Basel is still, after more than 30 years, used for electron diffraction in the courses of the Department of Physics of the University of Basel, though in a somewhat simplified form (without case).

What has happened to the other five microscopes installed in Switzerland is unknown. No details could be found about the fate of the more than 25 TTC microscopes installed abroad as well.

Related to the Swiss contributions on electron optics, the electron diffraction apparatus should also be mentioned. The development of this apparatus by the Trüb, Täuber & Cie. AG practically came parallel to the electron microscopes (see chapter by E.Kellenberger). The production and supply of these units (ca. 70), in various designs, was done at the beginning by TTC and later (e.g. Eldigraph KD4 & KD-G2) up to 1970 by Balzers AG in Liechtenstein.

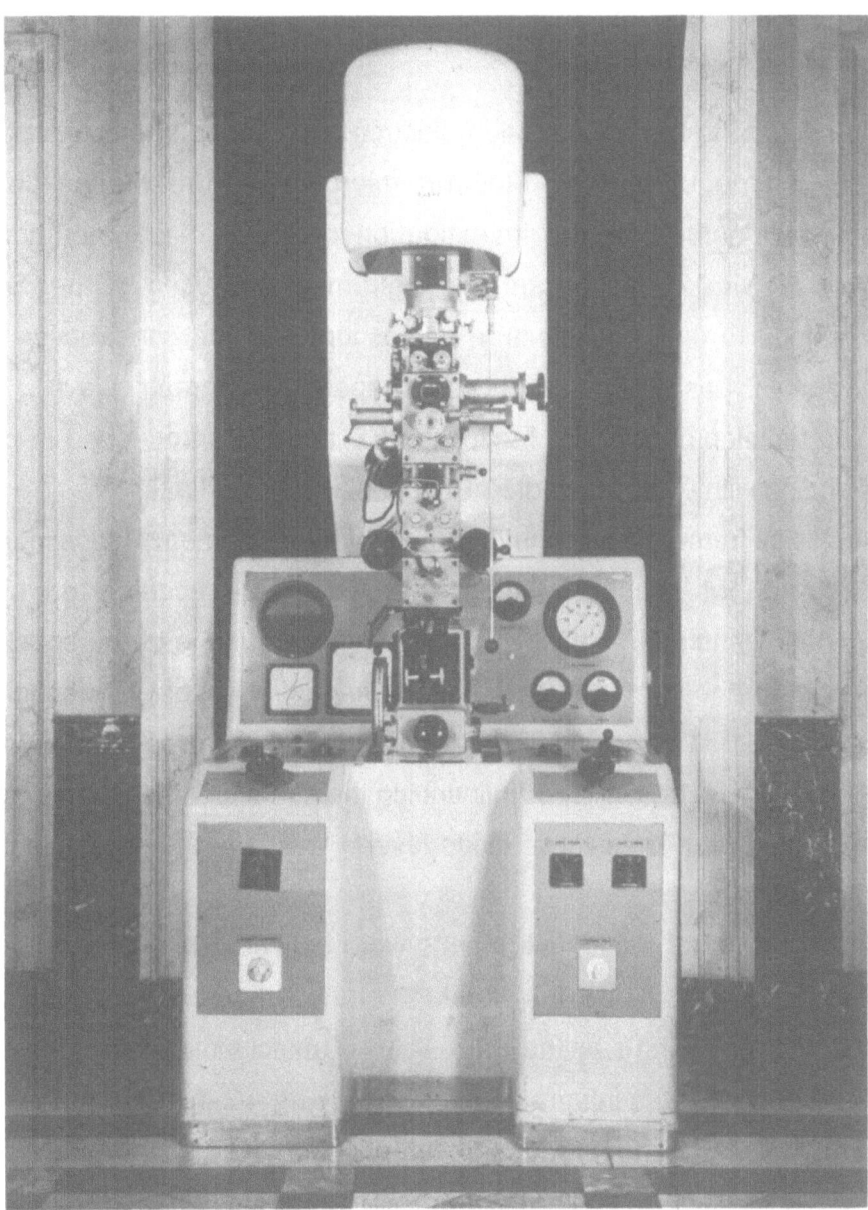

Fig. 6:   "Babar", constructed from the two TTC microscopes in Geneva.
(Photograph: Musée d´art et d´histoire, Geneva)

The Emission Electron Microscope

In contrast to the transmission electron microscope, the development of a new type of surface electron microscope, the Metioskop KE (commercial name) for the investigation of metallic and nonmetallic surfaces (7) was continued and the instrument produced. (Other Swiss contributions to emission electron microscopy by the construction of a prototype by E.Bas will be described in a separate paragraph below).

The development of the Metioskop KE and KE2 proceeded in close collaboration with G.Möllenstedt in Tübingen, FRG (8) (Fig. 7).

In this instrument, ionic bombardment of the specimen surface induces the emission of secondary electrons, which are accelerated in an electrostatic immersion objective and a magnified image of the specimen surface is produced by means of electromagnetic lenses. The specimen temperature could be varied in the Metioskop between room temperature and 1200 °C. The possibility of influencing the surface of the sample by ion beam etching (e.g. removal of oxide layers) was a further example of its versatility (9, 10).

A total of 20 of these instruments were delivered between 1961 and 1967 to customers all over the world. 13 of these had been manufactured by TTC, the remaining 7 by the Balzers AG. The scientific staff of the Trüb, Täuber & Cie. AG was integrated into the Balzers AG in 1965. At this time Balzers AG started to produce the Metioskop KE3, a photoemission electron microscope, under L.Wegmann following ideas by E.Ruska, W.Engel and G.Möllenstedt (Fig. 8).

Fig. 7:   Emission electron microscope Metioskop KE.

The instrument was designed for the investigation of solid surfaces.

For the emission of photoelectrons, the specimen surface was irradiated with ultraviolet light from high pressure mercury lamps. The light, collected by a quartz lens system, was focussed onto the specimen surface by an anode acting as a mirror. The acceleration of the low energy photoelectrons occured in an electrical field of about 80-100 kV/cm between the specimen surface and the anode. After passage of the accelerated electrons through the anode aperture, the desired magnification was adjusted by three electromagnetic lenses.

The magnified image of the specimen could be observed on a fluorescent screen and by removing this, it could be photographed.

The excellent contrast conditions allowed a continuous investigation over the entire temperature range from room temperature to 2000 ^{o}C both in photoemission and in thermionic emission. Details about the construction and some applications can be found in (11, 12, 13, 14). In the meantime, 10 of these instruments had been manufactured and delivered to customers by the Balzers company. The development and production of electron microscopes was stopped completely by the Balzers AG in 1974.

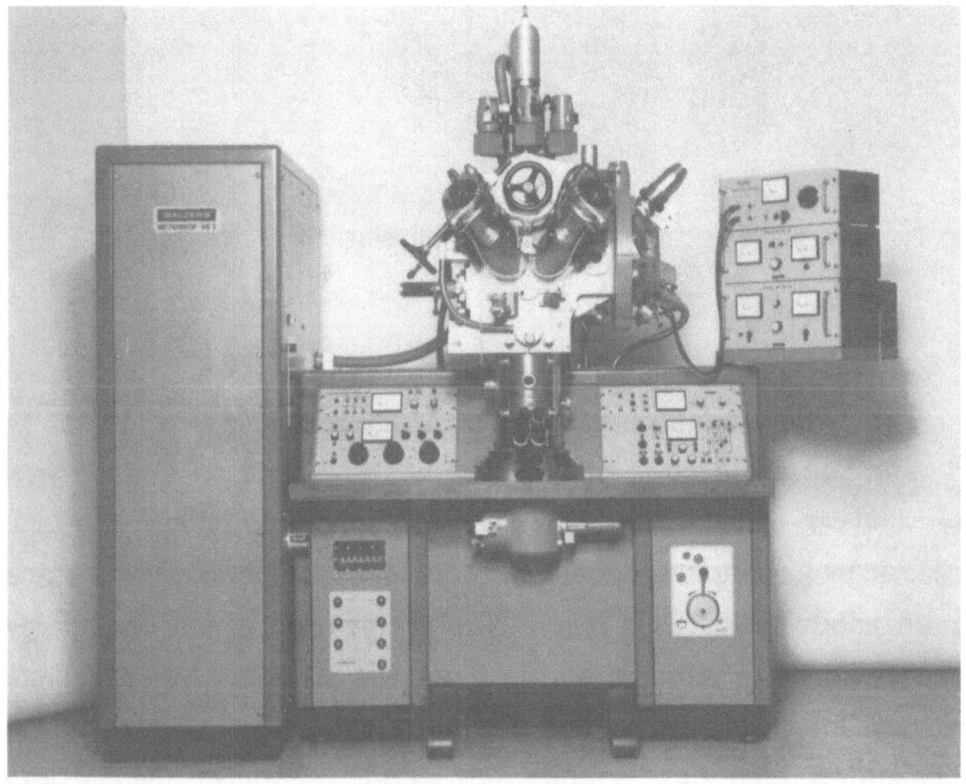

Fig. 8:   Balzers Metioskop KE3.

Technical Description of the TTC Transmission Electron Microscopes

The essential peculiarity of the TTC electron microscope has certainly been its cold cathode, introduced by Dufour in 1914 in his oscillograph and further developed by G.Induni (15) for application in the electron microscope.The electron gun with cathode and anode, constructed as a gas discharge tube, made it possible to regulate the required beam current by variation of the pressure through an air inlet valve. This system has been used by the TTC company, although in a strongly modified version (6) from the first test microscope to the last instrument delivered to a customer (Ecole polytechnique, Lausanne, J.-P.Borel) in 1963. Detailed values for the properties of the cold cathode were investigated by L.Wegmann and M.Gribi and have been published (16).

Two electrostatic lenses (objective and projective) have been conserved over the whole period as well. The reason for this was the possibility to compensate an eventual variation of the high voltage by coupling the voltages of the cathode and of the electrostatic lenses and thus avoiding a possible influence on the imaging quality. The high voltage could be selected in steps and amounted to 50 kV for the instruments up to the type KM4 and to 70 kV for the successor model KM5.

The Imaging System

The first test electron microscope was equipped with an electromagnetic condensor, an electromagnetic objective, and an electromagnetic projective lens. In 1945 the scheme of an electrostatic/electromagnetic lens system was published (see Fig. 2).

The prototype of a further developed instrument was presented in 1946 and had the following lens arrangement: Electromagnetic condensor lens, electrostatic objective lens, electrostatic imaging lens and electromagnetic projective lens. It was characteristic that this arrangement was considerably changed in the delivery of the first electron microscopes designed as KM. So, e.g., the instrument of the ETH Zürich had an electromagnetic condensor lens, an electrostatic objective lens and an electromagnetic projective lens (17).

A lens arrangement practically identical to that of the prototype of 1946 was used in 1952 in a new series of instruments named KM4 (5).

In contrast to the instruments of the type KM with fixed lens positions, in the new instruments of the KM4 series, both the electrostatic imaging and electromagnetic projective lens could be swiveled from the ray path. This made it possible to image the specimen with the electrostatic objective lens only, making a screening magnification of 120x possible. By switching off the voltage of the objective, an aberration free diffraction pattern could be obtained. By swiveling in and switching on the various lenses, the following magnification ranges were obtained: 800 to 1400x, 2500x, 8000 to a maximum of 14000x. For the centering of the optical system, the swiveling lenses could be easily positioned from the outside. The resolution of the KM4 electron microscope was specified as 2-3 nm.

In the latest transmission microscopes of the type KM5 produced in series by the TTC company with newly developed lenses, the following arrangement was used: an electromagnetic condensor lens, an electrostatic objective lens, an electromagnetic intermediate lens and an electrostatic projective lens (Fig. 9). This arrangement produced a broad continuous range of magnifications up to 35000x with minimal image distortion. The focal length of the electrostatic lenses was about 6 mm. The field strength of these electrostatic lenses lay at about 200´000 V/cm.

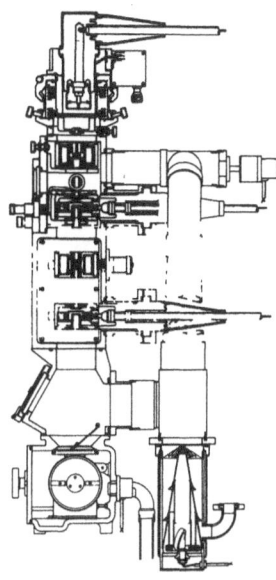

Fig. 9:   Microscope column and part of the vacuum system of the KM5 instrument, Basel.

Figures 10 and 11 show such an electrostatic lens of the KM5 in complete and partly disassembled state.

The use of an imaging system with a combination of electrostatic and electromagnetic lenses (Fig. 12) had a number of advantages:

1.   By switching on one or both of the electrostatic lenses fixed steps of magnification with only little image distortion were obtained (e.g. in the KM5 70x and 1800x).

Fig. 10: Complete electrostatic projective lens of the KM5, Basel.
Fig. 11: Partly disassembled electrostatic lens of the KM5, Basel.

Fig. 12: Set of lenses of the KM5: electromagnetic (top) and electrostatic (bottom) lenses.

2. The combination of electrostatic objective and electromagnetic imaging lenses gave a continuous range of magnifications from 800 to 3000x.

3. Adding the electrostatic projective lens enlarged the continuous range of magnification from 4000 to 35000x.

The resolving power of a KM 5 electron microscope was specified as 2 nm by the manufacturers. The physical principle and the action mode of electrostatic and electromagnetic lenses have been described in detail by E.B.Bas in 1960 (18).

The Vacuum System

The vacuum system designed by G.Induni in the construction of the first test electron microscope was very simple and consisted essentially of a prevacuumpump VV2 (manufactured by Micafil) and a molecular pump type HPI in the construction mode Trüb, Täuber - Holweck, which had been developed already earlier for the high voltage cathode ray oscillograph. The pumping rate of this relatively slowly rotating molecular pump (ca. 3000 rpm) was about 15 l/sec at a pressure of 10^{-3}Torr. The molecular pump yielded an oil vapour free vacuum and was used by TTC until 1956 in the construction of electron microscopes.

When the high voltage was raised from 50 to 70 kV, it had to be recognized that the pumping rate of the molecular pump was too low to guarantee the safety for operating the instrument.

For the KM5 electron microscopes, therefore a more powerful, water cooled oil diffusion pump type VD 200 with an "untransparent" baffle was constructed by TTC, which exhibited a pumping rate of 100 l/sec when the baffle was introduced. As prevacuumpump a DUO5 (5 m^3/h) of the Balzers company was used. By introducing a buffer tank with a large volume, the prepump could temporarily be switched off during microscopic work. The vacuum system was controlled by electromagnetic valves. For vacuum control, ionisation sensors and Pirani gauges with their corresponding measuring instruments, all own products of TTC, were used.

Whereas the vacuum during operation of the earlier TTC microscopes was in the range of 10^{-3} Torr, it was improved to 10^{-4} - 10^{-5} Torr in the 70kV instrument KM5.

The High Voltage Supply

The high voltage transformer for 50 kV (50 Hz) as well as the heating transformer had to be placed at some distance from the TTC electron microscopes because of their electromagnetic stray fields.

The alternating voltage conducted to the instrument through a cable was transformed into direct voltage via a rectifier tube, capacitors having a high capacity and resistors and then fed to the cold cathode and to the electrostatic lenses.

The supply of the KM5 instruments with their 70 kV high voltage occured by means of a high voltage transformer for 70 kV. The direct voltage was supplied from a rectifier with capacitor and filter chain placed in an oil tank. The high voltage transformers for the electron microscopes were produced by TTC. The rectifiers for 70 kV were supplied by E.Haefely & Cie. AG in Birsfelden.

The Photographic System

The very first "Übermikroskop" built by G.Induni was equipped with an external camera for photographing from the fluorescent screen. Already for the prototype presented in 1945, it was planned to install a cassette for double plates (9 x 12 cm) or a film cassette (9 cm) which could be exchanged together with the drying medium without flooding the microscope tube. The electron microscopes of the KM series (e.g. the ETH instrument) also contained an airlock between the camera space and the electron microscope tube and could thus be kept evacuated when the roll film cassette (12 images) was changed.

The KM4 series had the same system; however an exchangeable cassette for 12 plates (6.5 x 9 cm) was available.

The only difference in the KM5 series was the presence of two exchange cassettes for 3 or 10 plates (6.5 x 9 cm), respectively.

The Development of the Ultramicrotome

The first prototype of a Swiss ultramicrotome was constructed in 1949/50 at the University of Geneva by D.Danon and E.Kellenberger (Fig. 13). According to a publication by these authors (19) a section thickness of 0.04 µm could be obtained. The instrument was equipped with a mechanical advancement and could be used with a metal knife (specially sharpened razor blade) or with a glass knife. The construction principle was based on a spindle which shifted an inclined plane vertically; the resulting movement was reduced by a lever system onto the advance axis. On the spindle was a latch wheel of large dimension which was transported by one or several teeth according to desired section thickness. This microtome was operated manually by means of a crank disk. An improved version of this construction, in which the specimen was swiveled backwards after each cutting process and therefore could no longer touch the knife edge was described by E.Kellenberger in 1956 (20).

Fig. 13: Prototype ultramicrotome by Danon and Kellenberger, 1950.

Of the version published in 1950 (19), a small number was produced by the Trüb, Täuber & Cie. AG under the type designation DM2 (21) (Fig. 14).

Production of the improved version of 1956 was disregarded, as in the meantime an excellent construction (the Sorvall MT-1 microtome (22)) had been introduced and dominated the market.

As is well known, the diamond knife had been introduced in 1953 by H.Fernández-Morán. From the early to middle sixties, a number of former coworkers of the IVNIC (Instituto Venezolano de Neurologia e Investigaciones Cerebrales) had tried to become independent, i.e. to produce diamond knifes under their own names in Europe and the USA.

Fig. 14: TTC ultramicrotome DM2, 1952.

Therefore a new situation arose for the ultramicrotome users:

More or less independent from the specific manufacturer, the quality of the delivered diamond knives from serial production was not constant, i.e. they varied considerably in their cutting performance.

Only through the efforts of the company Diatome AG, Biel, founded in 1970, which improved the quality and perfectioned the production methods, it became possible to have diamond knives with a constant and excellent quality available today.

24

The Vacuum Evaporators

Under the guidance of L.Wegmann, the Trüb, Täuber & Cie. AG also constructed a small vacuum evaporator in 1957, which was called VAA3 and a few of these were produced (Fig. 15).

Fig. 15: Vacuum evaporator of the Trüb, Täuber & Cie. AG.

This unit allowed to perform all of the customary evaporation work (replica and shadowing techniques) with carbon, platinum, gold, silicon oxide, etc..

By means of the arched shape of the current supply leads and the adjustable holder of the heating tungsten filaments, baskets or foils, the specimens which were positioned on a rotatable table could be shadowed under practically any angle. A 2kV feedthrough enabled the glowing of samples for influencing the surface charge or the preparation of C-layers by means of an inlet for benzene vapour.

The pumping system consisted of a baffled oil diffusion pump type VD 200 (100 l/sec) and a prevacuumpump of the type Balzers DUO5.

All sensors and gauges for the vacuum control, as well as an efficient transformer (24 V/80 A) and various accessories such as air inlet valves etc. were produced by TTC. The only exception was the coupled vacuum valve block, produced by the Micafil company.

Further production of this evaporator unit had unfortunately to be given up regardless of its high versatility and considerable efficiency. Due to the high production costs, it could not compete with the products of foreign suppliers. At least one such instrument was in constant use until the late seventies in the former laboratory for electron microscopy of the University of Basel.

Acknowledgements: The author thanks the Balzers AG and especially Mr. R.Graber for support and many valuable discussions.

The Vacuum Evaporators of the Companies Balzers AG and Balzers Union AG. Principality of Liechtenstein

It was soon realized that electron microscopical specimens prepared by shadowing or evaporation techniques contributed a lot to the knowledge of the ultrastructure of the most various objects. The development of the required vacuum evaporators and units was mainly advanced by the company Balzers AG, partly in close cooperation with university institutes (e.g. ETH Zürich) and their performance was constantly adapted to the latest requirements.

Their universal applicability, the high quality in production and the simple and comfortable handling of these plants led to a world wide use.

The Balzers AG company, located in the Principality of Liechtenstein, was founded in the intention to produce ultrathin layers on glass surfaces for optical applications by evaporation of suitable substrates in an ultrahigh vacuum. For this purpose a high vacuum evaporation unit had first to be developed and produced. The result of this development was named BA 500. A first unit was sold by Balzers in 1950 to the company Kern AG, Aarau. Only shortly after this, a second unit was delivered to the Institute for General Botany at the ETH in Zürich (A.Frey-Wyssling, K.Mühlethaler) for the preparation of thin layers for electron microscopical preparation. This instrument (Fig. 16) was later used as basis for the freeze etching units constructed by H.Moor et al. in 1961 (23), in which for the first time a microtome was located inside a vacuum evaporation instrument.

Based on the experiences in the field of the evaporation technique collected by Balzers in the meantime, it was decided to construct a first serially produced electron microscopic preparation plant named BA 350E (Fig. 17) and thus the origin of a new product line of the company's sales program was founded. This type was offered in the years 1958-1961.

After this the successor model BA 350 G was introduced and marketed until 1971. The deep freezing chamber integrated in this type made it possible to shadow and coat specimens at low temperatures down to -150 $^\circ$C as well as to prepare freeze drying specimens. In the years 1971-1979, a new freeze etching plant under the name of BAF 301 (D or T for diffusion or turbomolecular pump) was produced, in which all manipulations within the vacuum chamber were no longer done by lifting the steel vessel, but by opening a large door on the front side. This model was later replaced by the BAF 400, which is still in the program today.

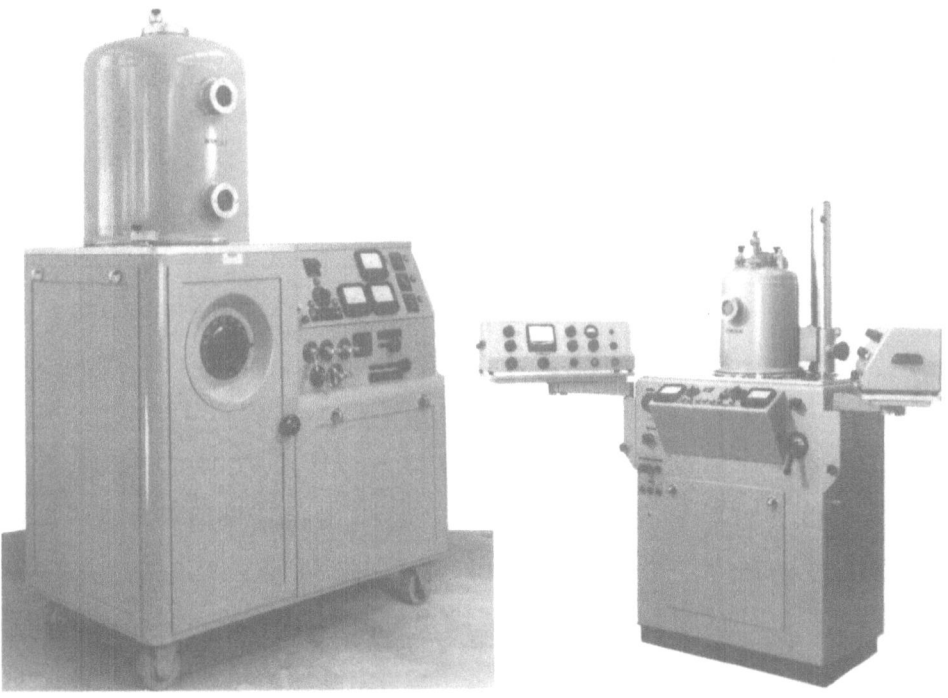

Fig. 16: Vacuum evaporator BA 500.
Fig. 17: Electron microscopic preparation unit BA 350E.

Many of the problems occurring in electron microscopic preparation could be solved with the small evaporating unit Mikro BA 3, a very reasonably priced apparatus available from 1961-1973. Its production was ceased in 1974 and it was replaced by the model BAE 080 D. Since 1978 the version BAE 080 T was available, which included a newly developed turbomolecular pump developed by the company A.Pfeiffer Vakuumtechnik Wetzlar GmbH (F.R.G.) for the production of a vacuum as clean (oilfree) as possible. Further experience with this pumping system was gained from 1970 to 1979 with the small evaporating unit BAE 120. The multiple possibilities of these small evaporating units were based on the use of a T- or cross-shaped recipient constructed from pyrex glass (as in the older Mikro BA 3) and an elaborate exchangeable flange system.

In 1978 a very powerful high vacuum evaporation unit, the BAE 370, was introduced, including according to the customers specifications a diffusion pumping system (BAE 370 D) or a turbomolecular pumping system (BAE 370 T). The specific properties of this unit were mainly the use of a large glass recipient and of a fully automatic pumping control. The concept and space requirements of this unit corresponded essentially to those of the classical evaporation units as offered also by competitors.

Following a wide spread wish of the users to a reduction in size of such apparatus, an intense development of the table top model MED 010 was started in 1982 and serial production of this small high vacuum coating unit was already started in 1983 (Fig. 18).

This instrument built in an unit construction system made most of the usual coating techniques used for transmission and scanning electron microscopy possible by means of a simple exchange of the required flanges. The so far latest model in the series, the BAE 250 Turbo, appeared in 1985 and shows a further increase in performance thanks to the larger diameter and height of its recipient and to a more powerful turbomolecular pump (Fig. 19).

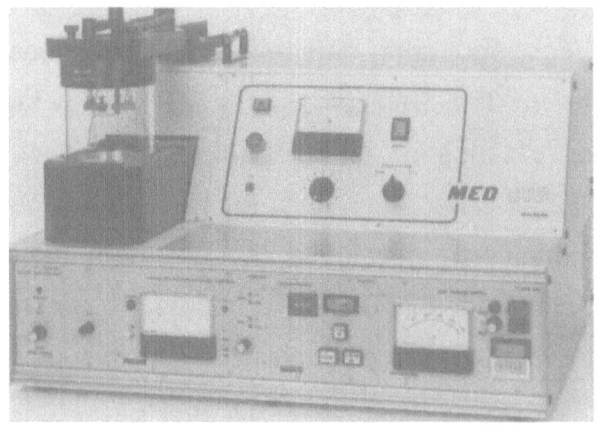

Fig. 18: Small high vacuum coating unit MED 010 (diameter of recipient 108 mm).

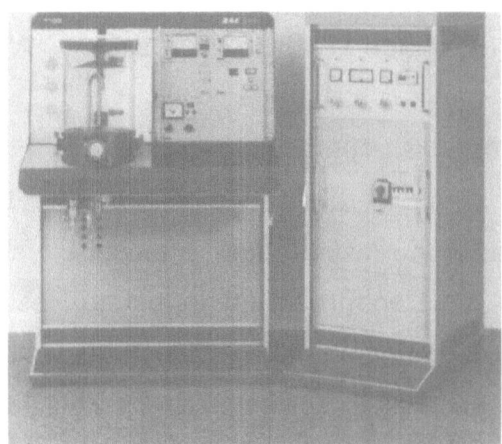

Fig. 19: High vacuum coating unit BAE 250 Turbo with glass or metal recipient, diameter 250 mm.

The great popularity of the Balzers evaporation units is certainly based on the large number of available accessories, as e.g. electron beam evaporator sources, sputtering equipment, film thickness monitors, etc. (see catalogues of the Balzers Union AG*), allowing reliable and controllable performance of all desirable kinds of vacuum coating.

(* Since 1982 all evaporation units for electron microscopic purposes and their accessories were developed, manufactured and delivered by the Balzers Union AG, founded in 1972 as a daughter company of the Balzers AG especially for handling the electron microscopic specimen preparation program).

Acknowledgements: The author thanks the Balzers Union AG and Mr. S. Böhler for support and supply of figures.

Additional Contributions to the Swiss Development of Electron Beam Sources, an Emission Electron Microscope and to High-Voltage Electron Microscopy (HVEM)

This review of Swiss contributions to electron microscopy would be incomplete without short references to the long term research and development work of the Institute for Technical Physics (former director: E.Baumann) and the Department for Industrial Research (AFIF) of the Federal Institute of Technology (ETH) Zürich, as well as of the company E.Haefely & Cie. AG, Basel. The Institute for Technical Physics and the AFIF, the latter founded in 1937 by initiative of F.Fischer, were involved for many years in the research and development of electron beam techniques and vacuum electronics (24). From the large variety of publications of these laboratories, the present author has selected a few contributions dealing with electron beam sources, electron optics and electron microscopy in general. To make electron microscopical experiments possible at all, an electron optical bench in a high vacuum system had been constructed in 1951 (25). By its means a number of problems occurring in the development of electron sources and imaging systems, as e.g. the sputtering caused by ion bombardment etc. could be studied in detail and solved. In this context, a so-called electron-gun with ion separation was constructed already in 1952, which could later on be improved in several respects, e.g. lifetime. This so-called bolt cathode was an indirectly heated tungsten cathode for electron guns (e.g. 26, 27, 28) especially developed for the use in the cathode ray tube of the television large screen projection system (Eidophor-system), but was later on used as well as an object in an emission electron microscope (29).

The knowledge and experiences acquired by means of the electron optical bench (30) proved to be a good basis for the development of a versatile emission electron microscope (Fig. 20) featuring imaging with both thermal and secondary electrons (31).

The two stage imaging system of this instrument comprised an electrostatic immersion objective of the Brüche-Johannson type (32), the electron optical data of which had been evaluated numerically-theoretically and experimentally in dependence of the most important lens parameters (33). The second stage of the imaging system consisted of an electrostatic projective lens (31). The development of a heatable, rotatable and tiltable object stage (34) greatly enlarged the investigation possibilities of this emission electron microscope. Further details about the electron optics of this instrument as well as some examples of applications (as e.g. surface

Fig. 20: Swiss emission electron microscope (AFIF, 1956).

imaging of metal specimens at high temperatures) are included in (35).

The principal differences between the electron microscopical imaging methods (by transmission and emission electron microscopes) and their physical reasons have been pointed out in a review article (36).

Applications of the electron beam technique at AFIF were thereon not limited to the television large screen projection system and to emission electron microscopy, but extended to further developments, such as the electron beam welding machine, the electron beam zone melting plant, the 90 kV X-ray tube for dental diagnostics, the high temperature Auger electron spectrometer and other physical apparatus (see review article 24). In the last years, the research staff of the AFIF was, amongst other projects, involved in developing the hybrid electron gun together with the cylindrical mirror energy analyzer (CMA), Auger electron spectroscopy (AES) and ion scattering spectroscopy (ISS) for elaborate surface analyses (37, 38, 39).

Towards the end of the fifties, scientists of various branches (physicists, metallurgists, biologists, physicians, etc.) pointed out that the accelerating voltage for transmission electron microscopy (at that time usually 100 kV) should be drastically increased. Some of the essential reasons for this requirement will be pointed out below. The increased energy of e.g. 1´000´000 Volt for electrons offers a higher penetration power due to the reduced electron scattering cross section, thus the investigation of e.g. etched or thin polished metallic specimens or biological objects embedded in resins, the so-called thin sections, becomes possible. Increasing the beam voltage decreases the chromatic effects improving the imaging and therefore leads to the expectation of a higher resolution at a given specimen thickness. Thin sections (about 0.5 to several μm thick) imaged at different tilt angles relative to the beam axis yield a three-dimensional impression of the structural relations within cells when evaluated stereoscopically. The possibility to study or even image living bacteria in an aqueous medium enclosed in a microchamber between two thin collodium foils also appeared to be fascinating. Several review articles (40, 41, 42, 43, 44) treat the problems, developments, applications and results of HVEM.

In connection with the project of a so-called high voltage electron microscope with an accelerating voltage of 1.5 MV in the "Laboratoire d´Optique Electronique" in Toulouse, France (then under the direction of G.Dupouy), a close collaboration with AFIF, Zürich, and the company E.Haefely & Cie. AG, Basel, was started in 1957/58. This led to the joint development of an electron accelerator for 1.5 MV (45, 46) finding international appreciation. A special electron gun (Fernfokus-Elektronenkanone) (47) developed at AFIF was used as electron source. For the principles of construction of this accelerator, the injection system, the arrangement of the accelerator steps, etc., see (45). The company E.Haefely & Cie. AG in Basel had a long term experience in the construction of high voltage installations (transformers, rectifiers, etc.) and of the problems involved with insulation (electrical feedthroughs up to 760 kV). They took over the task to produce the high voltage source for 1.5 MV (from transformers to symmetrical cascade generators), but also its voltage stabilization (better than 10^{-5} during approx. 3 min). The complex technical problems connected to this task were not unknown to the E.Haefely & Cie. AG, since they had already developed and delivered a direct voltage supply for 300 kV for the first ion accelerator built at the Department of Physics of the University of Basel during the second world war. Already in 1948 the delivery of a neutron generator for 1200 kV to the same institute followed. The intense interest of the nuclear physicists in such accelerators had the consequence that a number of other neutron generators (from about 300 kV to 1200 kV) were delivered world-wide in the fifties. Based on the acquired experience, even a 4 MV cascade accelerator plant (ion accelerator) was developed jointly with the Department of Physics of the University of Basel in 1959 and put into operation (48, 49).

These experiences were the best preconditions for the construction of the 1.5 MV electron accelerator for the Toulouse project. The ordered

Fig. 21:   1.5 MV electron accelerator for the Toulouse electron microscope.

accelerator plant could indeed be delivered in 1959 and its operation
started in 1960 (Fig. 21). The high voltage electron microscope itself was
developed and built by the physicists and technicians of the Toulouse
institute. The interesting results obtained with this instrument, e.g. in the
investigation of metallic specimens and other solids led to the delivery of a
similiarly constructed 750 kV accelerator to the Cavendish Laboratory of
the University of Cambridge, England, (then under the directorate of
V.E.Cosslett) in 1964 and even of a 1 MV plant to the US Steel Corporation

in Monroeville, Pa., U.S.A. for the operation of an RCA electron microscope in 1967. Its external features are almost identical to that shown in Fig. 21. The disadvantages of this open (under external atmospheric pressure) accelerator construction were its enormous space requirements (distance from the cathode to the electron microscope about 6 m !) and the occurence of certain electrical interference problems. These problems were practically eliminated through the development of a pressurized gas insulation for the accelerators for electron microscopes (50), the construction of which was based on earlier cooperations with the Department of Physics of the University of Basel in the development of a pressure tank ion accelerator (48, 49). Accelerator plants for electron microscopes in the so-called twin pressure tank design (51) of E.Haefely & Cie. AG became known worldwide and about a dozen of these could be delivered for use with the commercially available 1 MV electron microscopes type EM7 of the AEI Scientific Apparatus Ltd., GB (Fig. 22).

According to E.Haefely & Cie. AG, about 40 ion, neutron or electron accelerators (high voltage, etc., according to customer's requirements) were manufactured and delivered between 1948 and 1979. High voltage electron microscopes of various manufacturers are today operating at 1.2 MV or more in a sufficient number for the requirements of the research laboratories of industry and universities. Therefore, since 1982 E.Haefely & Cie. AG have concentrated in this field mainly on the construction of a 300 kV high voltage supply for the Philips electron microscopes EM 430 and CM30, and about 100 of these have been delivered up to now.

For additional information about Swiss contributions to the instrumental developments of electron microscopy, refer to the review articles (52, 53).

Fig. 22: Electron accelerator of the twin pressure tank type for AEI
electron microscopes.

Acknowledgements: The author thanks E.Haefely & Cie. AG and Mr.
R. Gleyvod for information and for the permission to reproduce Figures 21
and 22. Special thanks go to Mr. J.R.Günter for arrangement and translation
of this chapter.

References

1. G. Induni, Neue Zürcher Zeitung, nr. 796, 10 May 1944.
2. G. Induni, Neue Zürcher Zeitung, nr. 357, 28 February 1945.
3. G. Induni, Vierteljahresschrift d.Naturforsch.Ges.Zürich **90**, 181 (1945).
4. L. Wegmann, Trans.Instr.and Meas.Conf., Stockholm, 73 (1952).
5. L. Wegmann, Neue Zürcher Zeitung, nr. 1562, 16 July 1952.
6. M. Gribi, M. Thürkauf, W. Villiger and L. Wegmann, Optik **16,** 65 (1959).
7. L. Wegmann, Chem. Rundschau **14**, 499 (1961).
8. G. Möllenstedt and H. Düker, Optik **10**, 152 (1953).
9. F. Jobin, M. Gribi and L. Wegmann, Schweiz.Archiv f.angew. Wiss. u. Technik **27**, 453 (1961).
10. L. Wegmann, Schweiz.Archiv f.angew.Wiss.u.Technik, **30**, 143 (1964).
11. W.Engel,Proc.6th Intern.Congr.for El.Microsc., Kyoto, Vol. **I**, 217 (1966).
12. R. Graber, M. Gribi and L. Wegmann, Proc. 4th Europ. Reg. Conf. on El. Microsc., Rome, Vol. **I**, 111 (1968).
13. L. Wegmann, Proc. 5th Int. Congr. on X-ray Optics and Microanal., Tübingen, 356 (1968).
14. L. Wegmann, Handb. d. zerstörungsfreien Materialprüfung, Oldenbourg München 1969, R. **31**, 1 (engl. translation: Balzers High Vacuum Report nr. 23, September 1969).
15. G. Induni, Helv. Phys. Acta **20**, 463.(1947)
16. L. Wegmann and M. Gribi, Proc. Stockholm Conf. on El. Microsc., 41 (1956).
17. L. Wegmann, Optik **7**, 263 (1950).
18. E.B. Bas, Schweiz. Archiv f. angew. Wiss. u. Technik **26**,1 (1960).
19. D. Danon and E. Kellenberger, Archives des Sciences **3**, 160 (1950).
20. E. Kellenberger, Experientia **12**, 282 (1956).
21. L. Wegmann, Neue Zürcher Zeitung, nr. 318, 13 February 1952.
22. K.R. Porter and J. Blum, Anat. Rec. **117**, 685 (1953).
23. H. Moor, K. Mühlethaler, H. Waldner and A. Frey-Wyssling, J.Biophys. Biochem.Cytol. **10**, 1 (1961).
24. E.B. Bas, AFIF Jubiläumsschrift, 49 (1977).
25. E.B. Bas, Z. angew. Physik **6**, 404 (1954).
26. E.B. Bas, Ph.D. Thesis nr. 1813, ETH Zürich (1949).

27. E.B. Bas, Z. angew. Physik **7**, 337 (1955).

28. E.B. Bas, Optik **12**, 377 (1955).

29. E.B. Bas, Z. angew. Math. Phys. ZAMP **8**, 203 (1957).

30. E.B. Bas, Neue Zürcher Zeitung, nr. 2679/80 (1956).

31. R. Aeschlimann and E.B. Bas, Proc. 5th Int. Congr. for El. Microsc. Philadelphia, Vol. **I**, D-9, (1962).

32. E. Brüche and H. Johannson, Naturwiss. **20**, 356 (1932); Annal.Phys. **15**, 145 (1932).

33. E.B. Bas and L. Preuss, Optik **21**, 261 (1964).

34. E.B. Bas, Proc. 6th Int. Congr. for El. Microsc., Kyoto, Vol.**I**, 223 (1966).

35. E.B. Bas, Le Vide **96**, 303 (1961).

36. E.B. Bas, Schweiz. Archiv f. angew. Wiss. u. Technik **26**, 1 (1960).

37. E.B. Bas and E. Gisler, Proc. 4th Int. Conf. Solid Surfaces, Cannes, Vol. **II**, 1108 (1980).

38. E.B. Bas, E. Gisler and F. Stucki, J.Phys. E: Sci. Instr. **17**, 405 (1984).

39. E. Gisler and E.B. Bas, Vacuum **36**, 715 (1986).

40. G. Dupouy, Adv. Optical & Electr. Micr. **2**, 167 (1968).

41. V.E. Cosslett, Q.Rev.Biophysics **2**, 95 (1969).

42. K. Hama, In: Advanced Techniques in Biological Electron Microscopy, J.K. Koehler (ed.), Springer Verlag, 275 (1973).

43. A.W. Agar, In: Practical Methods in Electron Microscopy, A.M. Glauert (ed.), Vol. **2**, Chapter 9, North-Holland Publ.Co., 297 (1974).

44. C. Humphreys, In: Principles and Techniques of Electron Microscopy, Biol. Appl., Vol.**6**, M.A. Hayat (ed.), Van Nostrand Reinhold Co. 1 (1976).

45. E.B. Bas, Proc Europ. Reg. Conf on El. Microsc., Delft, Vol.**I**, 126 (1960).

46. H. Adler, R. Minkner, G. Reinhold and J. Seitz, Proc. Europ. Reg. Conf. on El. Microsc., Delft, Vol. **I**, 122 (1960).

47. E.B. Bas and F. Gaydou, Z. angew. Physik **11**, 370 (1959).

48. R. Galli, E. Baumgartner and P. Huber, Helv. Phys. Acta **34**, 352 (1961).

49. J. Seitz, G. Reinhold and R. Minker, Helv. Phys. Acta **33**, 977 (1960).

50. G. Reinhold, J.L. Savary, K. Trümpy, H. Adler and J. Bill, Proc. 4th Europ.Reg. Conf. on El. Microsc., Rome, Vol. **I**, 27 (1968).

51. G. Reinhold and R. Gleyvod, IEEE Trans. on Nuclear Sci., Vol. **NS-20**, nr. 3, 378 (1973).

52. E.B. Bas, AGEN-Mitteilungen nr. 10, 120 (1969).

53. G. Reinhold, Die Weltwoche, nr. 47, 20 November 1970.

History of Electron Microscopy
in Switzerland
Edited by John R. Günter
© 1990 Birkhäuser Verlag Basel

THE SWISS STEM PROJECT

Andreas Engel
M.E. Müller-Institute for High-Resolution Electron Microscopy
at the Biocenter, University of Basel

Towards the end of the sixties, a new type of scanning electron microscope was developed by Albert Crewe and his graduate students in Chicago. Combining state-of-the-art electron optics with a field-emission gun, it was possible to demonstrate that a scanning microscope can achieve near atomic resolution. In addition, as outlined in a remarkable article by Crewe in 1970 (2), the scanning transmission electron microscope (STEM) should allow several novel imaging modes ideally suited for visualizing fragile biomacromolecules (for STEM imaging modes see Fig. 1). These revolutionary ideas and the first stunning dark field micrographs of unstained DNA molecules were presented in 1970 at the

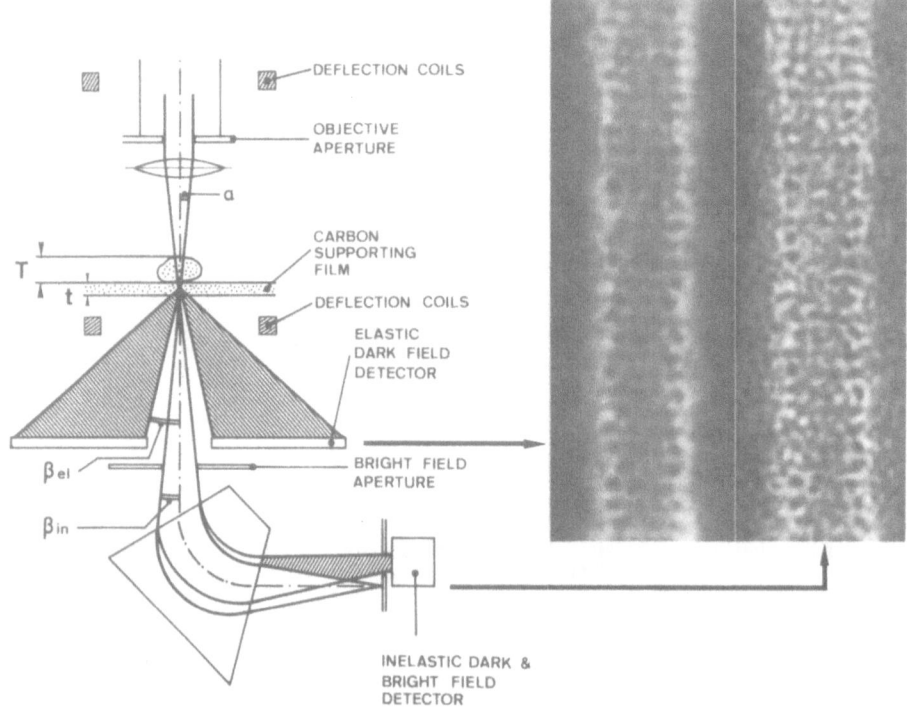

DEFLECTION COILS

OBJECTIVE
APERTURE

a

CARBON
SUPPORTING
FILM

T

t

DEFLECTION COILS

ELASTIC
DARK FIELD
DETECTOR

BRIGHT FIELD
APERTURE

β_{el}

β_{in}

INELASTIC DARK &
BRIGHT FIELD
DETECTOR

Fig.1: The high resolution objective lens focuses the electron beam of a STEM to a probe of a few Å diameter, while deflection coils move it along a raster allowing the specimen to be sampled at atomic resolution. Elastically scattered electrons diverging at large angles are collected by the annular detector and provide the elastic dark field signal. Near the optical axis, unscattered electrons are mixed with the low angle elastically scattered and the majority of inelastically scattered electrons. A spectrometer deflects those electrons which have lost energy at a larger angle than the unscattered electrons, thus facilitating the acquisition of the inelastic dark field signal. Unscattered and low angle elastically scattered electrons interfere to generate a coherent bright field signal which is collected through a small aperture placed on the optical axis. These sequential signals can be collected in parallel and processed on-line as the probe is scanned over the sample. The simultaneously acquired elastic dark field (DF) and bright field (BF) images of a negatively stained phage T4 giant tail are displayed with equal contrast to illustrate the different characteristics of these two imaging modes.

International Congress of Electron Microscopy in Grenoble. The enthusiasm induced by these results among the biologically oriented electron microscopists may be compared to the recent excitement fostered by the scanning tunneling microscope (STM) invented by Heinrich Rohrer and Gerd Binnig in Rüschlikon (1).

Shortly after the Grenoble Conference, Ewald Weibel, at that time chairing the Division for Biology and Medicine of the Swiss National Foundation for Scientific Research, invited the electron microscopists from Basel, Berne, Geneva and Zürich to a round table discussion on the possibility of setting up a STEM in Switzerland. Due to his initiative the Swiss STEM project was born and was, in fact, the first of its kind in Europe. The general idea was to install and operate a STEM in an environment of biologists, within a team providing both the technical background for the instrument as such, and for sample preparation and relevant biological projects. The task of exploring the potential of this new instrument for gathering structural information from biological samples was given to Eduard Kellenberger and his collaborators at the Biocenter in Basel. In his department, the conditions required for the STEM project were ideally met. Not only was there a stimulating environment for structural biology, but also the expertise in dark field microscopy of Jacques Dubochet was available, and an image processing group with Ross Smith and Ueli Aebi had just been started by Eduard Kellenberger. However, considering the particular technical aspects of Crewe's design such as a cold field emitter with a vacuum in the 10^{-11} Torr range and a digital data acquisition system, a physicist was required to nurse the STEM through its adolescence. At that time, it was not even clear whether the instrument should be built from scratch or whether a commercial unit should be acquired.

Just after finishing my thesis on holographic filters at the Institute for Applied Physics in Berne, I heard about the activities in Eduard Kellenberger's group and went to Basel to discuss the possible tasks for a young physicist. The STEM project as well as the intensive atmosphere at the Biocenter fascinated me and so I accepted Eduard's offer to join his group with enthusiasm. To learn the tricks of the trade, I went to Michael Beer at The John Hopkins University (JHU) where a STEM was close to being finished under the direction of Wendell Wiggins, a physicist with a background in high-energy physics and instrumentation. Over two fruitful years I learnt a great deal about the design of ultrahigh vacuum systems, modern analog and digital electronics and the just emerging techniques of real time digital data acquisition with minicomputers. Besides participating in the development of electronic gadgets and machining parts for the JHU STEM, I became fascinated by the image formation theory of STEM imaging modes which were also developed at that time in Chicago (21, 22). Application of the formalism used in Fourier optics allowed the various modes of coherent transmission electron microscopy to be simulated in considerable detail. This work led to the conclusion that the STEM annular dark field mode should be able to visualize single heavy atoms with a distinctness no other electron imaging mode could provide (5). This was, in fact, practically demonstrated by Joseph Wall, Michael Isaacson and John Langmore (11, 18), all graduate students with Albert Crewe, as well as by The John Hopkins University STEM once it became fully operational.

Back in Basel, somewhat different goals were set as it became clear that biological material could never withstand the electron dose required to image single atoms. We planned to use the high contrast and electron collection efficiency of the STEM dark field modes to record low-dose images of fragile biological samples. After a thorough evaluation, the only dedicated commercial STEM available in 1975, the British Vacuum Generators HB5 was ordered. In the years to follow this instrument, equipped with an excellent cold field-emission gun operating at 100 kV, a high quality electron optical column and a flexible scan system proved the best possible choice for our applications. In addition, it has been a great experience to collaborate with Ian Wardell, Peter Bovey and John Colling of VG Microscopes Ltd, who designed the HB5. Several new developments resulted from this collaboration, the most important one for our work being an annular dark field detector that counted each electron it collected.

The STEM HB5 was delivered early in 1976 and was immediately used to demonstrate its potential for visualizing negatively or positively stained samples with unprecedented clarity (6). It was due to the expertise of Jacques Dubochet, acquired during many frustrating hours with conventional dark field microscopy, that we could record convincing micrographs from negatively stained tobacco mosaic virus (TMV) or positively stained DNA (Fig. 2). As documented with several samples, it was not only possible to record micrographs matching the best obtained by conventional bright field electron microscopy (6,9), but also to acquire dose series permitting the radiation damage to be directly visualized (Fig. 3). Subsequent experiments revealed that even the best STEM micrographs from negatively stained samples did not lead to a breakthrough, as it was the preparation method rather than the microscope that created the stringent limitations. Besides this, such samples proved to be sufficiently stable under the electron beam so that the low dose capability of the STEM was not absolutely required. Compared to the conventional transmission

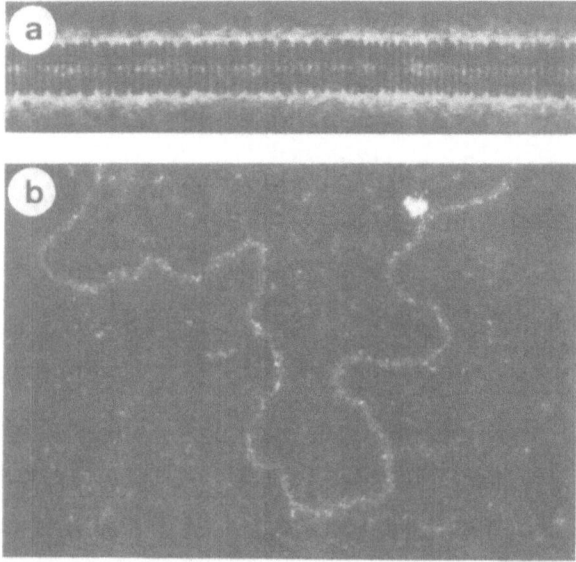

Fig. 2: Tobacco mosaic virus (TMV) subunits assemble into stacks of paired disks, each disk exhibiting a thickness of 23 Å. This pairing is obvious on the elastic dark field STEM micrograph displayed in (a) and is found to be occasionally disrupted by disk triplets. DNA becomes positively stained by uranyl ions which form clusters of various sizes depicted as bright spots in (b). Compared to the approximately 20 Å diameter of the DNA double helix, the size of uranyl clusters indicates an instrumental resolution of 5Å.

electron microscope the STEM turned out to be rather difficult to operate. While the former is quite tolerant with respect to focus adjustment and allows the enhancement of a specific spatial frequency range by choosing the appropriate amount of underfocus, the STEM needs to be tuned close to the Scherzer focus for optimum transfer of structural information. Furthermore, as the STEM transmits low frequencies unattenuated, specimen staining must be completely uniform.

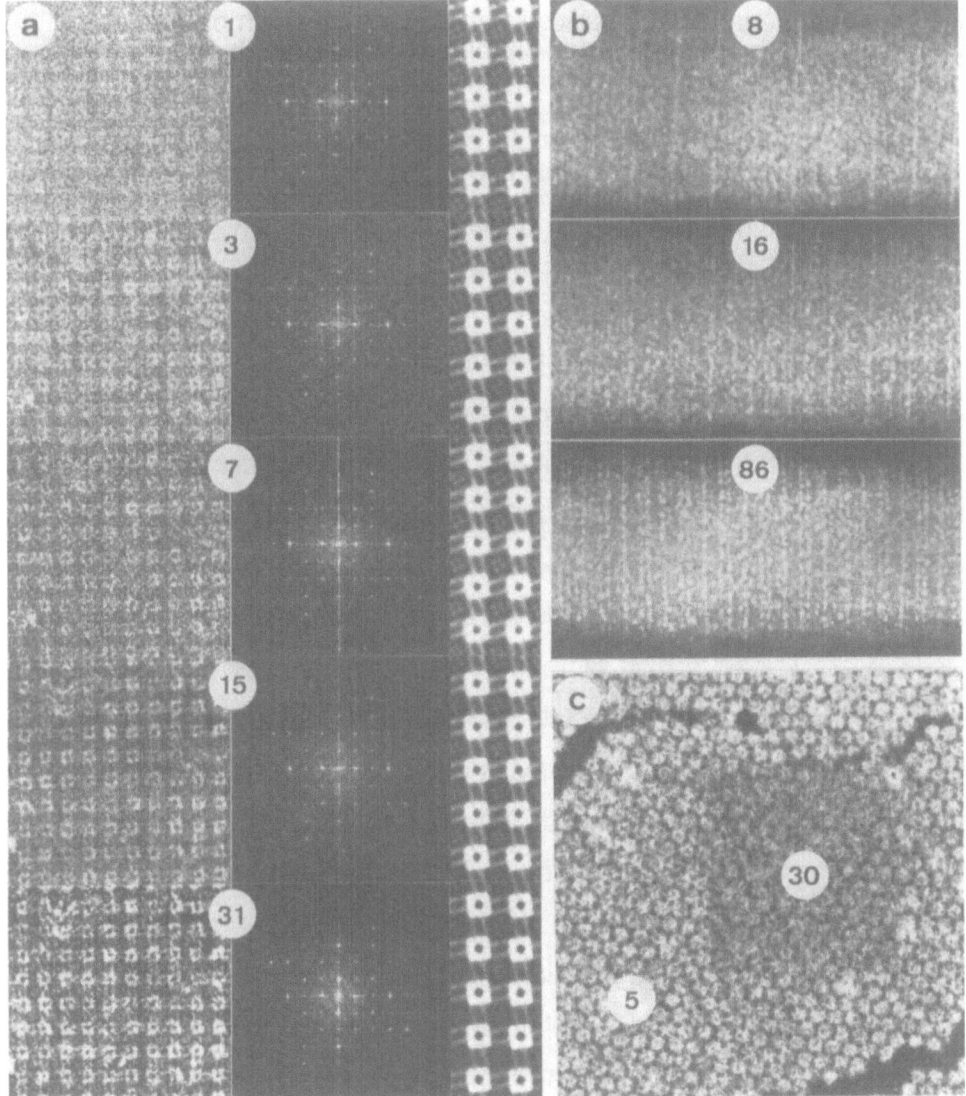

Fig. 3: Dose series unveil the sensitivity of differently prepared samples towards the electron beam. The series recorded from a negatively stained bacterial surface layer (T-layer) displayed in (a) illustrates the remarkable stability of such samples. No obvious structural rearrangement is discernible on the diffraction patterns or averaged T-layer unit cells over the dose range from 1 to 31 electrons/Å² (a; right side), whereas the quantum noise is

abundant in the micrograph taken at 1 electron/Å² (a; top left). A dose series of a specifically labeled tropomyosin paracrystal illustrates beam induced shrinkage and a dramatic rearrangement of the beam sensitive tetra-mercury label (b). Mass-loss is demonstrated by the central square of a freeze-dried, unstained hexagonally packed intermediate (HPI) layer of *Deinococcus radiodurans* exposed to 30 electrons/Å² (c). The center-to-center distances of repeating units are 130 Å in (a), 400 Å in (b), and 180 Å in (c), while the numbers represent the accumulated dose.

So it became clear to us that the full potential of the STEM could not be fully exploited with negatively stained samples.

In 1977 our custom-made on-line digital data acquisition system was finished and thoroughly tested. We then looked for applications in which the possibility of counting electrons with considerable precision could be used to extract biologically relevant information. As the elastic scattering cross-sections are known for all elements, it is a trivial task to calculate the number of atoms within a biomacromolecule by measuring the fraction of electrons it scatters. This kind of mass determination was first proposed by Elmar Zeitler in the early sixties (20) - at that time a revolutionary idea indeed. It was clear to us that the STEM was the best instrument for such measurements. The recording dose could be determined with high precision, and 70% of the elastically scattered electrons were collected, counted and stored on-line in digital format. There are two prerequisites for such experiments. Firstly, the preparation of unstained samples free of residual salts but with the biomacromolecules still in their native oligomeric arrangement. Secondly, quantitative evaluation of the digitally recorded dark-field images: this requires a software package facilitating integration of counts over regions of interest in the digitized images. All this was developed over the following years (7), tested with suitable standards such as TMV or fd-phage, and applied to a considerable variety of biological structures, a procedure which is being continually refined and used today with many important biological specimens (Fig. 4). STEM mass determination provides so many obvious advantages that it is astonishing in how few laboratories such measurements are routinely done. (i) Minute amounts of material are required as a single good grid allows thousands of molecules to be measured. (ii) Several kinds of structures can be measured simultaneously provided that they can be distinguished by their size and shape. (iii) Spherical oligomers can be evaluated as well as filamentous structures (mass-per-length) or sheet-like assemblies (mass-per-area). (iv) An enormous range of molecular weights can be analyzed. The mass-per-length of a DNA double-helix (2,000 Da nm^{-1}) is just measurable, whereas the total mass of a T4 bacteriophage head ($2 \cdot 10^8$ Da) can easily be determined. Last but not least, mass maps can be determined with surprising resolution, provided the sample exhibits a periodic arrangement of subunits (8). In addition, due to the speed of today's computers the analysis is quick and mass data can be extracted with a relatively small amount of work. Today STEM mass determinations are routinely performed by the group of Joseph Wall in Brookhaven (19), at the EMBL in Heidelberg and at the Biocenter in Basel.

Eduard Kellenberger's aim remained to use the STEM as an imaging instrument, taking advantage of the dark field mode for recording micrographs of unstained sections. Together with his collaborators, Erich Carlemalm, Werner Villiger and Jean-Dominique Acetarin, he initiated low-denaturation embedding procedures during the late seventies leading to new types of embedding resins (4, 12). In spite of the greatly improved structural preservation demonstrated by thin sections of various test

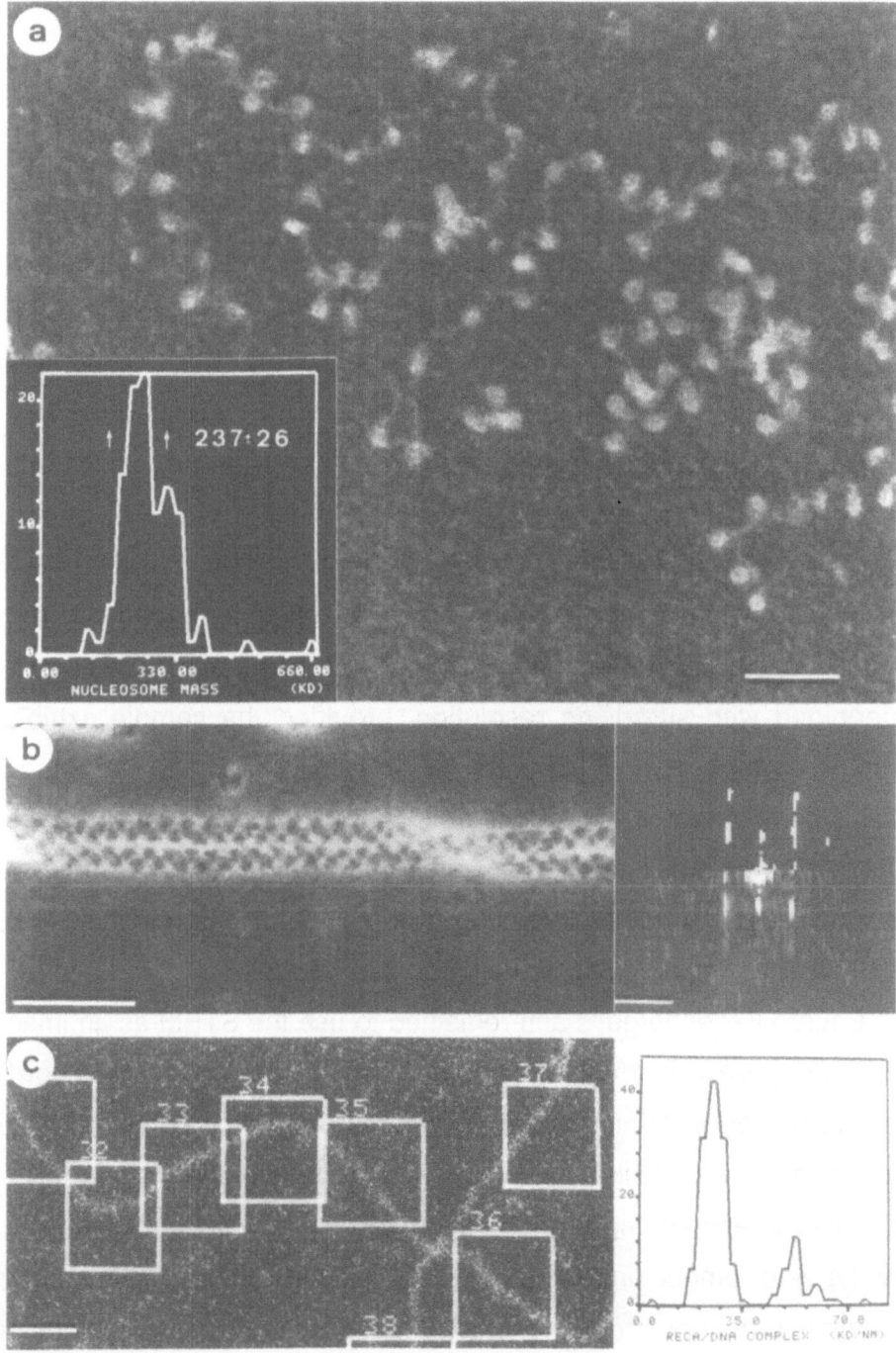

Fig. 4:   The mass within a sample element irradiated by the scanning
            probe can be determined from the ratio of scattered to impinging
            electrons measured with high precision in the STEM. As these
            numbers are transferred to a computer and can be reconstituted
            to an image, the mass of a biomacromolecule or the mass-per-
            length of a filamentous supramolecular aggregate can be
            calculated by simple integration over these structures. The
            calibration factor is either measured using a standard of known
            mass or calculated from scattering theory. Chromatin prepared by
            spreading with a surfactant reveals nucleosomes connected by
            double-stranded DNA (a). Nucleosome masses peak at a value
            corresponding to one histone octamer plus 146 base-pairs, but
            the shoulder towards higher mass values indicates that a distinct
            structure must be associated with a significant fraction of
            nucleosomes (histogram in (a)). Complexes formed by recA and
            DNA exhibit a helical structure which is visible after negative
            staining (b) and reflected by layer lines in the diffraction pattern (b;
            right side). No substructure is discernible on unstained recA-DNA
            fibers (c), but their mass-per-length allowed the number of recA
            monomers per helical repeat to be calculated with surprising
            precision. Scale bars represent 500 Å .

samples after staining, the corresponding unstained sections did not exhibit the expected quality when imaged in the STEM elastic dark field mode. Collection of the inelastically scattered electrons by a spectrometer in one channel and the elastically scattered electrons via the annular detector in the other channel (Fig. 1), enabled the crispness of the image to be increased by normalization of the elastic with the inelastic image (i.e., "ratio"-image). This particular dark-field mode, introduced by Albert Crewe as "atomic number" contrast (2, 13) produced beautiful micrographs from unstained septate junction sections (Fig. 5) (3, 10). However, several experimental difficulties prevented the routine application of this imaging mode and there were contrast inversion phenomena on ratio-contrast micrographs of thick sections (9) that called for a quantitative explanation. Also, it was not quite clear what mechanisms were primarily responsible for the sharpening effect of the ratio-contrast imaging mode. It was clear,

52

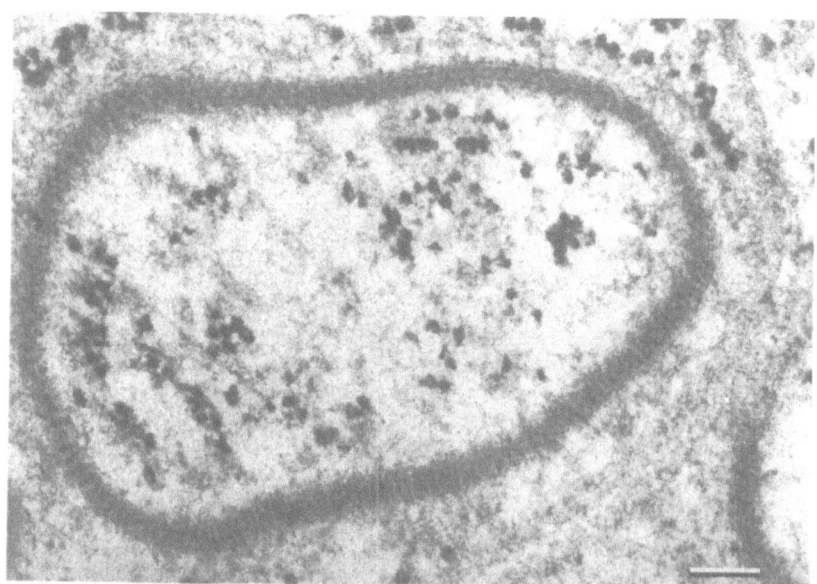

Fig. 5: Septate junctions are special structures consisting of regularly packed, juxtaposed transmembrane channels used by the cells to exchange their cytosol. A ratio-contrast image of an unstained thin Lowicryl HM20 section from such a junction obtained by dividing the elastic dark field signal by the inelastic dark field signal of the STEM (see Fig. 1) unveils the transmembrane nature of the proteinous channels in the lower part of the micrograph, whereas the two adjacent membranes are clearly visible in the upper part. The scale bar represents 500 Å. (By courtesy of E. Carlemalm and E. Kellenberger).

however, that multiple scattering must occur in sections with a thickness close to or even exceeding the mean free path between scattering events. To study these effects in detail, a Monte Carlo simulation was programmed by Rudolf Reichelt who joined the STEM group in 1982 to design a high resolution ß-spectrometer. Although it was evident that the fraction of multiple scattering events could be exactly calculated by Poisson statistics, and that there were ways to approach the problem analytically, we chose the numerical simulation because it allowed experimental data to be taken into account elegantly and the simulation to be displayed graphically in a straightforward manner. This Monte Carlo program package allowed the systematic exploration of ratio-contrast imaging for various materials. Its limitations could thus be determined and nonlinearities such as contrast reversal could be quantitatively explained (14). As expected, multiple scattering was indeed responsible for the loss of resolution in the annular dark field mode; electrons collected by the annular detector that were inelastically scattered at least once carried the delocalization effect of inelastic events into the elastic dark field channel (16). This artifact is eliminated by normalization of the annular detector signal with the inelastic dark field signal collected by the spectrometer. Last but not least, these numerical simulations also provided a basis for calibrating the concentration measurement of resin-embedded protein or DNA in thin sections (15).

The Swiss STEM project is not history. It is now integrated into the M.E. Müller-Institute for High-Resolution Electron Microscopy at the Biocenter in Basel founded in 1986. This Institute, which is a unique joint venture between the private industry and the State of Basel, became possible due to a generous endowment made by the Maurice E. Müller Foundation, Biel. Ueli Aebi, the director of this Institute was one of the few who realized the enormous potential of quantitative STEM for biological structure research and thus fostered further methodological and instrumental development. We anticipate to use the Basel-STEM not only for mass-determination or mass mapping-techniques now routinely used in our Institute (17) - but also to introduce quantitative determination of element composition at high spatial resolution. Ultimately this kind of element-mapping will find the same broad application in structural biology as this has been the case for mass-determination.

Acknowledgements

I would like to thank Eduard Kellenberger for his continuous support during many fruitful years, to Jacques Dubochet, who introduced me to electron microscopy of biomacromolecules, and to my collaborators Rudolf Reichelt, Robert Häring and Robert Wyss, who continuously helped to refine the methodology and instrumentation for quantitative STEM. I am grateful to all the colleagues who have shown interest in the progress of the Swiss STEM and have become engaged in collaborative projects that took advantage of the possibilities provided by this sophisticated instrument. Major contributions concern studies of the recA-DNA complex from *E. coli* (Elisabeth DiCapua and Theo Koller), the HPI-layer of *Deinococcus radiodurans* (Wolfgang Baumeister), the reaction center of *E. halochloris* (Harald Engelhardt), human keratin filaments (Ueli Aebi), myosin (Doris Walzthöni and Theo Wallimann), and the phage T4 head stoichiometry (Werner Baschong). I also would like to thank Shirley Müller for her help with this manuscript. The STEM project has been supported during many years by the Swiss National Foundation for Scientific Research through grants to Eduard Kellenberger and A.E., by the Department of Education of the Kanton Basel-Stadt, and more recently, by the M.E. Müller-Foundation of Switzerland.

References

1. G. Binnig and H. Rohrer, Helv. Phys. Acta **55**, 726 (1982).
2. A.V. Crewe, Quart. Rev. Biophys. **3**, 137 (1970).
3. E. Carlemalm and E. Kellenberger, EMBO J. **1**, 63 (1982).
4. E. Carlemalm, R.M. Garavito and W. Villiger, J. Microscopy **126**, 123 (1982).
5. A. Engel, J.W. Wiggins, and D.C. Woodruff, J. Appl. Phys. **45**, 2739 (1974).
6. A. Engel, J. Dubochet and E. Kellenberger, J. Ultrastruct. Res. **57**, 322 (1976).
7. A. Engel, Ultramicroscopy **3**, 273 (1978).
8. A. Engel, W. Baumeister and W.O. Saxton, Proc. Natl. Acad. Sci. USA **79**, 4050 (1982).
9. A. Engel and R. Reichelt, J. Ultrastruct. Res. **88**, 105 (1984).
10. R.M. Garavito, E. Carlemalm, C. Colliex and W. Villiger, J. Ultrastruct. Res. **80**, 344 (1982).
11. M. Isaacson, J. Langmore and J.S. Wall, IITRI/SEM/**1974** ,19 (1974).
12. E. Kellenberger, E. Carlemalm, W. Villiger, J. Roth and R.M. Garavito, Low Denaturation Embedding for Electron Microscopy of Thin Sections. Publ. by Chemische Werke Lowi GmbH, D-8264 Waldkraiburg, p. 1 (1980).
13. E. Kellenberger, E. Carlemalm, W. Villiger, M. Wurtz, C. Mory and Ch. Colliex, Annals N.Y. Acad. Sci. **483**, 202 (1986)
14. R. Reichelt and A. Engel, Ultramicroscopy **13**, 279 (1984).
15. R. Reichelt, E. Carlemalm, W. Villiger and A. Engel, Ultramicroscopy **16**, 69 (1985).
16. R. Reichelt and A. Engel, Ultramicroscopy **19**, 43 (1986).
17. R. Reichelt, A. Holzenburg, E.L. Buhle Jr., M. Jarnik, A. Engel and U. Aebi, J. Cell Biol. **110**, 883 (1990).
18. J.S. Wall, Chemica Scripta **14**, 271 (1979).
19. J.S. Wall, IITRI/SEM/**1979** /II , 291(1979).
20. E. Zeitler and G.F. Bahr, J. Appl. Phys. **33**, 847 (1962).
21. E. Zeitler and M.G.R. Thomson, Optik **31**, 258 (1970).
22. E. Zeitler and M.G.R. Thomson, Optik **31**, 359 (1970).

II. THE PIONEERS

History of Electron Microscopy
in Switzerland
Edited by John R. Günter
© 1990 Birkhäuser Verlag Basel

THE BEGINNING OF
ELECTRON MICROSCOPY IN ZÜRICH

Kurt Mühlethaler
Schönisteig 1
8303 Bassersdorf

As a student in the 5th semester at the ETH Zürich in the Department of Biology, I had to complete a course in General Botany during the winter term 1941/42. Prof. A.Frey-Wyssling, Head of the Institute for General Botany, lectured at that time on the topic: "Sublightmicroscopic Morphology of Plant Cells". This fascinating introduction into the world of macromolecules and their arrangements in cell walls, cytoplasm, nuclei etc. was the beginning of my interest into this field. Therefore I decided to make my diploma work in the next semester under his guidance. At the beginning of February I went to his office to discuss the theme of my diploma work. He began our conversation saying: "A few minutes ago the engineer of Trüb, Täuber, Mr. G.Induni, was here. He told me that they have built a new electron microscope and for the preparation of specimens, they are looking for a biologist. This is perhaps of interest to you and in case you decide to join his group, you could as well carry out your diploma work there on cellulose structure in plant fibers. Since the winter term is over, you can immediately start working there."

So far I had not read or heard a thing on electron microscopes, because during the war we had only a few months for studying and for the rest of the year we had to join the army service. As Frey-Wyssling pointed out to me the electron microscope should open up a new world in biology and therefore I decided to join Induni's group. In order to learn more about this instrument and its applications, I went to Wipkingen next day, where I asked for G.Induni. When he welcomed me in his office, I was immediately impressed by his charme and his overwhelming enthusiasm for his new microscope. He showed me his laboratory with the electron microscope (Fig. 1) and explained to me its construction. Deeply impressed by the new apparatus, I agreed to join his group. This team was rather small and consisted only of the designing technician F.Häusler and the mechanician A.Hugentobler. These three men did build the first instrument in only few weeks during the winter 1941/42. With great pride Induni showed me his first photograph of asbestos fibers which he took on the 6th of January 1942 as shown in Fig. 2. To get his picture he had simply dusted the fibers onto the screen of the object holder. The magnification was only 1500 times.

The Trüb, Täuber company was a well known factory for the production of electric measuring instruments, such as Volt- and Ampère-meters. Together with the BBC in Baden, Induni also constructed larger instruments, such as cathode ray oscillographs for high voltage

Fig. 1: Preparation room with the first electron microscope in the factory
of TTC in 1942.

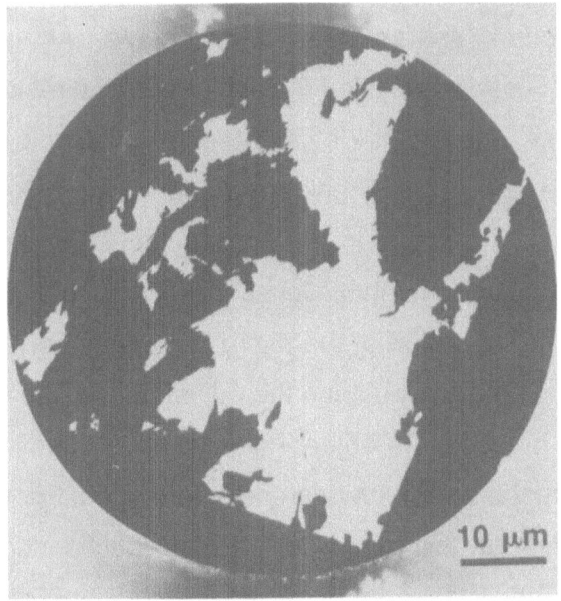

Fig.2: First electron micrograph taken by G.Induni on the 6th January
1942. Object: asbestos fibers.

measurements, a television scanning device for the Swiss National Exhibition in 1939 and an electron spectrograph for the Institute of Crystallography at the ETH in 1940. Therefore Induni had the required experience for building the electron microscope. As an electron source, Induni always used a cold cathode, first introduced by Dufour in 1922. This source was very simple and the gun consisted only of a discharge space between anode and cathode, where ionized gas particles were accelerated towards a metal plate to release the electrons. These were accelerated in the opposite direction towards the anode and by collision with the gas molecules, new ions were produced which were instrumental for a steady electron beam. For the regulation of the beam, a molecular valve was necessary for controlling the gas flow into the discharge space (see chapter by W. Villiger).

Also the vacuum system was very simple. As in all previous instruments, Induni used a so called molecular pump, directly flanged to the microscope column and in turn evacuated by a Micafil prepump. This system was stable against air breaks, needed no air-locks and no water cooling and was easy to maintain. The prototype, which I could use, was equipped with three electromagnetic lenses, which were supplied by a 12 V storage battery. The operation of the microscope was very easy, because the regulation of the cold cathode and the adjustment of the lens current did not require much practice. As object holders, small aluminium cylinders, covered by a fine metal screen on one end , were used. For the introduction of the specimen holder, the microscope had to be flooded and the front window in front of the stage removed. With a fork, the specimen holder could be removed and the new one introduced into the stage. For the photographic registration, a plate holder for two plates 9x12 cm was available under the viewing screen.

For the introduction into this new field of activity, Induni provided me with a few crumpled journals, where various papers by E. and H.Ruska, B.v.Borries, M.v.Ardenne and H.Mahl were published. From these publications, I could learn that the different properties of light and electrons make it inevitable that different kinds of preparations are needed for optical and electron microscopy. I learnt that the development of preparations suitable for examination under the electron beam is dominated by the extreme opacity of all matter to electrons. What is to be inspected must be very thin and of such a nature as not to be destroyed by the vacuum or the beam. The German authors postulated that for 50 kV electrons and preparations consisting mainly of light elements, a maximum thickness of the order of 0.2µm is needed. I realized that this requirement would cause a lot of difficulties and frustrations. From the available literature I learnt that the first step in specimen preparation starts with the supporting membrane. At this time, most authors recommended collodion as film material (1). The process for making thin collodion membranes was first introduced by Trenktrog in 1923 (2) and described in detail by v.Angerer in 1924 (3). The collodion was dissolved in amylacetate, and a few drops of this solution were given onto the surface of a water bath saturated with amylacetate. After allowing the film to harden for a couple of minutes, the membrane was mounted onto the grids, placed below, by lowering the water surface. After drying the films under a lamp, the grids were ready for mounting the specimens. The easiest objects for examination in the electron microscope were suspensions of small inorganic particles such as gold colloids, or evaporated metals such as beryllium or zinc. Particles suspended in water

64

were placed as minute drops onto the collodion covered grids by means of a small pipette capped with a rubber nipple.

After I had established a reliable method for supporting membranes, I was looking around for an interesting object. I learnt from a colleague that a well known diatome expert (Dr. F.Meister) had donated his collection to the ETH. This enabled me, together with my colleague R.Braun, to begin an investigation of various species. Electron micrographs of diatome shells had previously been published by Krause (4) and von Ardenne (5). Within a few weeks we could get numerous pictures of all kinds of diatomes. Fig. 3 shows such a silica structure of Diatoma vulgaris. Later on, some of our results were published (6) and Induni used a few of them for advertisements.

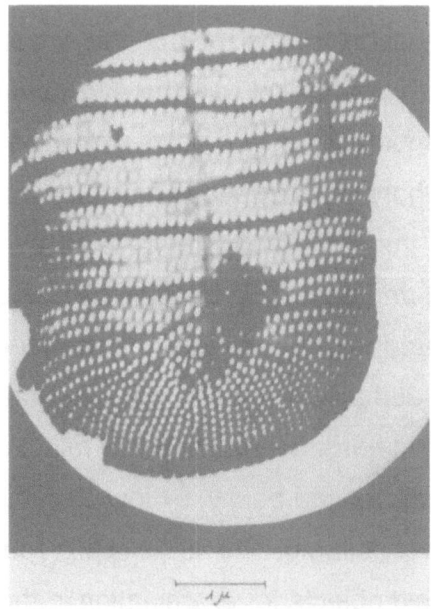

Fig. 3:   My first electron micrograph of a diatome (Diatoma vulgaris), 1942.

During the first phase in electron microscopy I had a lot of troubles, because the microscope was constantly under reconstruction. Every week new modifications were tried out, old parts were removed and new devices tested. These alterations reduced the working period to a few hours in a week. Also the exchange of grids and of the plates was very time consuming. A lock device was not available and therefore the whole column had to be flooded and subsequently the evacuation took more than 10 minutes before the microscope could be operated. In order to speed up the change of plates, phosphorous pentoxide was placed in a petri-dish behind the plate holder. Later the plates were replaced by a rollfilm box which allowed more photographs. This film box caused new problems, however, because the dry films became electrostatically charged and the discharges during unrolling caused lightning like figures on the film after developing.

Soon I became also aware of the fact that the lenses were not stable in their magnification. For calibration tests, I used shells of diatomes (Fig. 3) which at that time yielded reliable data concerning the magnification, alignment and lens defects. I noticed that the magnification decreased during the working period, because the battery became discharged. There were also constant problems with the pole pieces. Some of them were very astigmatic due to inhomogeneities of the metal and sometimes only a dirt caused a blurred image. Therefore Hugentobler was kept busy with the production of new pole pieces. In spite of these efforts to improve the quality of the electromagnetic lenses, no considerable improvement was achieved. The difficulties in producing quality pole pieces and the problems with the stabilization of the lens current led Induni to convert the next microscope to electrostatic lenses. He found that they were simpler to build and in addition they had the advantage that the same voltage could be supplied also to the electron source. Voltage fluctuations had also less influence on the image quality than the same fluctuations would have in the high voltage applied to an electromagnetic microscope.

The second microscope with electrostatic lenses was installed in the attic of the factory in a room which was extremely hot in summer and cold in winter. This new type of microscope caused to me even more troubles than did the previous one. The first handicap was the necessity of having to operate at fixed magnifications, which did not allow to enlarge an object detail to the best possible magnification for photography. Later on Induni accepted my complaints and introduced an electromagnetic focussing lens under the objective lens and an electromagnetic condensor. Also in the second microscope, Induni used a cold cathode and a molecular pump. As mentioned before, the cold cathode was only working in a poor vacuum and in order to emit the electrons, a steady air flow was adjusted by means of the valve. This gun, together with the low efficiency of the molecular pump, caused a number of problems in operating the electrostatic lenses. As soon as I raised the voltage to 40 kV, the lenses started to crackle, which was an indication of flashovers within the lens plates. Serious troubles were also caused by occasional traces of dust or other insulating materials in the bores of the lenses. They lowered the image quality or led to additional flashovers, which in turn damaged the lenses. These had then to be disassembled and repolished. Instead of making new pole pieces, Hugentobler was now kept busy with polishing lenses, which lasted one day delaying the microscope operation. Based on these rather sad experiences, I proposed Induni to return to electromagnetic lenses and to replace the gun by a heated filament in combination with an oil diffusion pump. He was shocked that I wanted to change his design and refused my proposal. He was convinced that the simplicity of his construction would bring a commercial success. He wanted to produce a microscope which should be completely insensitive to all external influences.

In contrast to the operation of the instrument, the preparation of the specimens turned out to be extremely difficult. Often I execrated my decision to join Induni's group. The preparation of particle suspensions such as diatome shells, colloids etc. caused no problems, but my diploma investigations did not yield any results. Prof. Frey-Wyssling was constantly asking for cellulose pictures, but I simply could not get any reliable image from the fibers I prepared. Everything I tried was useless. All preparations melted under the beam, and if they were transparent to the electron beam, only an unstructured mass remained. As described by M.v.Ardenne in his standard text book on electron microscopy (5), wedge-shaped sections could be obtained by cutting and could be examined along the thin edge of the wedge. I spent weeks cutting cellulose fibers thin enough for electron microscopic examinations, but without any success at all. My sections were completely untransparent for the electron beam. One morning, Induni brought his army pistol with ammunition and Hugentobler was asked to drill holes into the bullets, which I filled with cellulose fibers. A steel plate was used as target, and we expected that the high pressure during impact would dissect the fibers. After our funny shooting match, I tried to recover some of the fibers enclosed in the flat pressed bullets, but not the slightest traces of cellulose could be found.

The next preparation technique I tried was the ball-mill treatment, as described by Wergin (7). He used this method also for fiber preparations for electron microscopy and published a few micrographs, on which segmented elements which he named Dermatosomes could be seen.

68

According to his concept, a plant fiber should be composed of the Dermatosomes, enveloped in a noncellulosic matrix, like the cement in a brick wall. This view was controversial to Frey-Wyssling's model, which postulated a structure based on anastomosing long cellulose strands. Frey-Wyssling prepared a reply (8) to Wergin's publication (7) and questionned his results. According to Frey-Wyssling, the Dermatosomes could be artefacts caused by the treatment with the ball-mill. My own results with this preparation method did not yield enough evidence to prove this hypothesis. In contrary, I did not observe Dermatosomes as Wergin had described. At that time, K.Wuhrmann and A.Heuberger working at the Swiss Federal Institute for Testing of Materials in St.Gallen performed experiments on textile materials with ultrasonic waves (9). They observed disintegration of their objects during sonication. When I heard about theses experiments, I asked Heuberger to let me use his apparatus. As a result, I could show in the electron microscope that cellulose fibers are split up in fine fibrils if they are sonicated for 3 to 10 minutes (Fig. 4). For the first time I could show that Frey-Wyssling's concept of fiber structure seemed to be correct. My specimens yielded photographs showing bundles of cellulose molecules with various diameters, frequently as thin as 6 nm. Their length seemed to be limited within the fibers. These observations were in full agreement with Frey-Wyssling's concept of fiber structure.

Complementary to the sonicated fibers I studied also cellulose membranes secreted by Acetobacter xylinium and B. xylinoides cultured in liquid media. If the bacteria were grown on wort agar the colonies appeared as brown and slimy masses. This material, spread on a grid, contained numerous bacteria embedded in an amorphous substrate. In liquid cultures, thin cellulosic membranes at the surface of the medium are

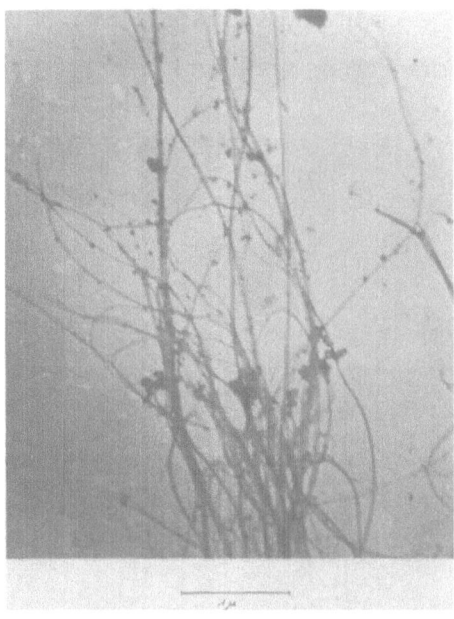

Fig. 4:   Hemp fiber after sonication.

secreted. For their preparation, a specimen holder was placed below the film and the culture liquid allowed to flow away. In this way, the thin cellulose film was spread onto the specimen holder film and could be examined (Fig.5). The threads we observed showed all the same diameter of about 20-30 nm. Their length could not be measured. Earlier X-ray diffraction patterns by Meyer and Mark (10) showed an orientation in threedimensional space, which made it possible to determine the critical unit of the cellulose pattern. Therefrom the authors concluded that the threads are crystalline. At this time the specimens had to be observed in the microscope without contrast enhancing staining or metal shadowing. Therefore the contrast of the microfibrils was always extremely weak. Focussing was a gamble and most of the photographs were out of focus. In

70

order to increase the contrast, we used repro-films, processed them in contrast enhancing developers and used extra hard paper for copying.

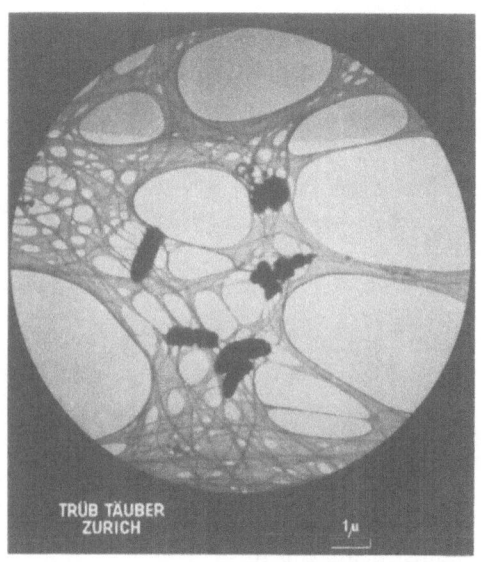

Fig. 5: Membrane of bacterial cellulose (B. xylinoides) one day after inoculation.

Similar cellulose fibrils as in membranes were also found in plant slimes of epidermal cells of quince, flax and hemp seeds (11). For preparation, a swollen seed was picked up with tweezers and carefully drawn over the collodion-covered screen, so as to leave behind a delicate film of slime. According to the results obtained from bacterial membranes and seed mucilages, the cellulose molecules seemed to form bundles of uniform diameters of about 20-30 nm, which we called microfibrils (12). These results did not quite confirm our earlier findings on sonicated fibers, where we had observed bundles of variable diameters. We therefore concluded that the microfibrils could be aligned to larger strands of variable sizes.

As time went by, I began to doubt if a technique for thin sectioning could ever be worked out. Therefore it seemed profitable to try replica methods to gain more informations of a wide variety of objects. Replicas prepared by pouring more or less diluted collodion over the surface in question and peeling the dried film from it were difficult to get. In porous material the collodion solution penetrated into the objects and could not be detached after drying. From glass or polished metal surfaces the collodion film could be recovered, but the contrast of the relief was very unsatisfactory. Therefore I did not spend more time on this technique.

In order to win prospective customers for his microscope, Induni was very eager to get images of various other objects, such as bacteria, viruses, plant and animal tissues, metallic alloys, paints etc. for his propaganda brochures. However, with the limited techniques which were available at this time I could not meet all his requirements. Nevertheless more and more clients showed up in our laboratory and most of them brought there own objects for investigation. All applicants became frustrated and uninterested if they were introduced into the preparation technique and the use of the microscope. One day also a movie producer from Basel, August Kern, came in and told us that he had got an order to shoot a film about the first Swiss electron microscope from the Wander company in Berne. Induni was very enthusiastic about this project, because he hoped the movie would make his instrument known to the general public. I was not pleased because I had to work on my thesis. Kern told me that the Wander company wanted influenza viruses to be shown in this film. They were interested in these bodies because the first pictures of these organisms had been published by a German group (13). I could foresee that this demand would cause a lot of problems to me, because nobody here had ever prepared these viruses for electron microscopy. First I went to the Institute for Microbiology at the medical school and asked for help, but they were not interested at all to spend time for such a project. They told me to ask the director of the Swiss Vaccine Institute in Berne for advice. He replied that they did not have this infectuous agent in culture, but would be ready to give me a starting suspension for inoculation of chicken embryos. In order to do these experiments I needed at least 50 eggs, which were not freely available during the war. Therefore I had to submit an application for eggs to the Swiss war-time economy office in Berne. At first the employee did not believe that this was a serious request and declined my application. I told Kern that I would be unable to prepare influenza viruses without eggs and that he and Wander should act in this matter. About a month later I received the eggs and inoculated them with viruses.

In the meantime, the movie based on a script which Induni and I had delivered to Kern was nearly finished and only the last scene with a close-up picture of the viruses seen on the fluorescent screen was missing. The camera operator Addy Lumpert with whom I made also the light microscopic pictures became very angry that he could not finish his film in time. In spite of my efforts to obtain convincing images of viruses, I could not get a single preparation in which particles of uniform diameters could be seen. All the eggs I opened after breeding looked completely normal and there was not a sign which indicated that the influenza virus had multiplied. From every part of the egg I made a swab but no result could be achieved and at the end I ran out of eggs. An ultracentrifuge with which I could separate the virus protein from the rest of the egg contents was not available. In this awful situation we decided to take a virus photograph from the German publication and from this picture I made a nice drawing in black and white which was subsequently projected onto the electron microscopic screen through one of the viewing windows. This screen was filmed through the front window and the movie was completed. At the film festival for scientific movies at Cannes, Kern received an award for his masterpiece and I got SFr. 300.-- for my collaboration during the whole summer term.

In the meantime the Trüb, Täuber company received the first order for a microscope from Prof. W.Feitknecht in Berne. This instrument was delivered in 1946 and H. Studer was put in charge of it. In the same year another instrument was sold to Geneva, where it was supervised by P. Dinichert and later by E. Kellenberger (see his chapter). Soon after these two microscopes were in operation, the first bad news for Induni came from Berne. Studer, unhappy with the fixed magnifications of the electrostatic lenses sawed up the microscope column to add a focusing electromagnetic lens. This operation caused a lot of problems because the additional lens could not be placed at the correct position between the object and the projecting lens. Also with Geneva poor Induni had problems, because Kellenberger was not satisfied with the quality of the lenses and tried all kinds of modifications to improve their performance (see chapter by Kellenberger on early work in Geneva).

Concerning myself, I was ready to leave the TTC laboratory at the end of 1947 because I had completed my thesis. From the Swiss-American Society for Scientific Exchange, I received a grant of 250 $ to spend one year in R.W.G.Wyckoff´s institute at the National Institutes of Health in Bethesda. At this time this laboratory was the top center for electron microscopy because Wyckoff, together with Williams, had introduced the technique for metal shadowing in 1944 (14) and his publications on phage development caused a sensation among scientists. When I arrived in

Bethesda in January 1948, three RCA microscopes of the EMU type were in operation. One was used by Wyckoff, the second by Scott and the third was reserved for me. From the manual I learned with the help of a dictionary how to operate the microscope. I realized very soon that this instrument was superiour to the TTC apparatus in every respect. After all the troubles I had had with this model the RCA construction was godsent.

The heated filament emitted a strong electron beam and even at the highest magnifications the image on the screen was bright and rich in contrast. The diffusion pump, backed by an efficient prepump evacuated the column within a minute and the electromagnetic lenses allowed a wide range of magnifications. The astigmatism in the lens could be eliminated by introducing iron in the form of small screws into the gap of the objective pole piece.

For shadow-casting, an RCA apparatus was available. Several metals were tested and due to the fact that chromium did not alloy with tungsten, this metal was generally used for evaporation. A small piece of it was placed in a tungsten basket which was made by winding a few centimeters of annealed tungsten wire around brass wood-screws as mandrils. The bell jar was large enough to place several grids around the source. The resulting layer was devoid of structure at ordinary magnifications and the film did not develop structure under the electron beam.

One day I saw an apparatus which was used for grinding various materials, e.g. to mince meat or prepare juices. It consisted of a motor casing and a cup with a built in propellar which could be mounted on the driving shaft of the motor. I got the idea of using this "Waring-blendor" for breaking up the cell walls and fibers. I tried it out, sedimented the heavier components and dried a drop of the fine suspension on the grid. After shadowing, I got the best pictures of cellulose walls I had ever seen before (Figures 6 & 7). With this simple technique I could image the structure of primary and secondary cell walls, as well as fibers of different origin. It was a breakthrough in cell wall research (15).

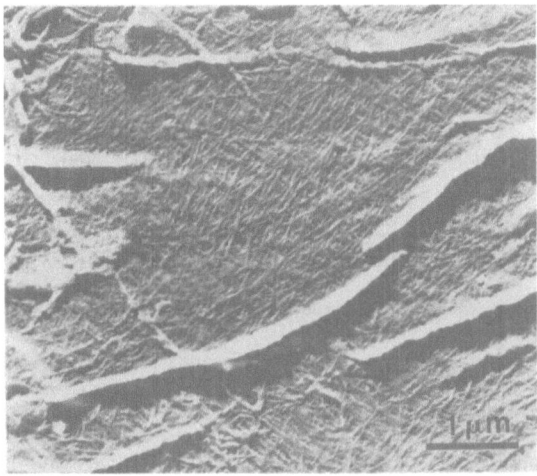

Fig. 6: Electron micrograph of the primary wall of a ramie fiber. For preparation, a Waring blender was used. Shadowed with chromium, 1948.

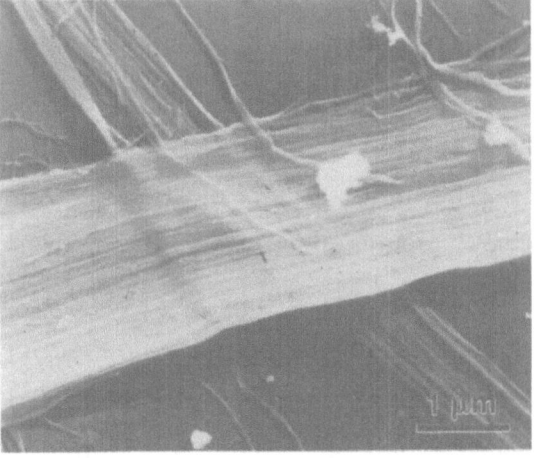

Fig. 7: Electron micrograph of the secondary wall of a ramie fiber.

At the same time I was also introduced into a new technique for the examination of surfaces. Scott, working with teeth, used a new replica technique which could be used for both optical and electron microscopical observations. He first made a negative replica with a cellulose-tape, called "Faxfilm". This tape was softened on one side with a suitable solvent, such as acetone and then pressed against the surface in question. The tape was thick and strong enough to be easily pulled from the surface. Subsequently this replica was covered with a silica film and the matrix dissolved.

Unfortunately I had to leave Wyckoff´s laboratory at the end of the year because the ETH had ordered a TTC microscope which was ready for delivery. I was appointed as research assistant and head of the new electron microscopy group. In my conditions of employment it was mentioned that I had to raise the necessary money for the laboratory expenses with service work for the industry. During my stay at Bethesda, I had asked Frey-Wyssling to cancel the contract with Trüb, Täuber and to buy an RCA microscope as well as a vacuum evaporator from the same manufacturer. However, he was not willing to do so and therefore the microscope was delivered at the end of 1948. No money was available at the ETH to buy a shadowing apparatus of the same type I had used at Bethesda. Therefore I decided to build such an instrument at the workshop of the institute. I copied the manual of the RCA instrument very carefully, photographed all the technical parts and documented myself with the engineering data. After my return to Zürich, I contacted the "Gerätebauanstalt Balzers" for the parts we needed for this construction. The chief engineer, Dr. Winkler, became very interested in our project and promised to construct such an apparatus according to the data I had collected. In 1950, the factory delivered the first evaporator of this model (see chapter on Instrumentation by W.Villiger).

After the new laboratory was set up, also the first candidate for a doctor's degree, Alfred Vogel, began to work in our group. Later he became the head of the electron microscope laboratory of the University Medical School.

In 1949, I received a new invitation from Wyckoff to return to Bethesda for another year. I accepted his offer with pleasure because important progress was achieved in the development of new sectioning techniques. I realized that a method allowing to produce thin sections routinely would bring a sensational break-through in biology and medicine. In 1948, Pease and Baker (16) published a sectioning technique using a modified Spencer microtome, by which sections of 0.1-0.2 μm thickness could be cut. One year later, Bretschneider (17) from the Zoological Laboratory at Utrecht published an article, in which he described his method. For cutting thin sections, he used an old type of rocking microtome which was originally designed by Horace Darwin in 1885. It owed the possibility of a minimal shift of the object towards the cutting blade to its typical transference of the screw movement to a lever, on which a hinch was fitted that turned a second lever, carrying the object holder. The same basic design was also used later on by Danon and Kellenberger (18) for the construction of their microtome which was subsequently manufactured by TTC. In the attic of our building, I found a similar type of rocking microtome which had been built in the last century by Jung in Heidelberg. For an introduction into the new thin sectioning technique I went to Utrecht, where Bretschneider showed me his method. He used OsO_4 and formalin for fixation in combination with chromic acid and potassium bichromate. As an embedding medium he used a mixture of two parts yellow beeswax and one part paraffin with a melting point of 72° C. For cutting he used an ordinary knife which was carefully polished on a hone and afterwards on leather. Cutting was done in a cold room at a temperature of 8° C. For this procedure he was dressed in a polar suit and looked like an eskimo. This was not exactly the field of activity I was looking for, but in spite of this handycap I tried it at home under the same conditions. The results were not very promising because I got very rarely a good section with a well preserved structure.

In October 1950 I arrived again in Bethesda where I was introduced into a new thin sectioning technique, worked out by Newman, Borysko and Swerdlow (19) from the National Bureau of Standards in Washington D.C.. They infiltrated their specimens after dehydration in alcohol with a mixture of methyl- and butyl-methacrylate instead of paraffin and polymerized this mass with 2,4-dichlorobenzoyl peroxide as catalyst at a temperature of 45° C for several hours. This was done in gelatine capsules which could be easily removed from the polymerized block in water. A Spencer microtome was used for thin sectioning in which the advance mechanism was replaced by a brass holder mounted at the specimen holder and which could be cooled by expanding carbon dioxide gas (Fig. 8). During the warming up period of the block about 50 sections could be cut. At first also metal knives were used, but in 1950 Latta and Hartmann (20) introduced glass knives which made sectioning much easier. When I stayed at Bethesda, also an experimental microtome made by RCA was delivered to Wyckoff. Basically it had a rotor that spinned without interfering vibration on a vertical axis, driven by the kind of air turbine employed in ultracentrifuges (21). Cutting and specimen collection took place in a closed chamber which could be evacuated and filled with dry air. The material to be sectioned was first chilled with dry ice or liquid air to preserve it from the damaging heat developed during high-speed cutting. Experiments we made with this device were not at all satisfactory, probably because no special knives for high-speed cutting were available.

Fig. 8: Spencer microtome adapted for thin sectioning (1950).

At my first stay in Bethesda in 1948 I learned the metal evaporation technique and during the second period 1950/51 the sectioning method.

When I returned to Zürich again, I could immediately introduce a reliable thin sectioning technique and therefore more biologists and physicians became interested in electron microscopy. Among the presumptive users the question arose as to whether live specimens could be observed in their original structural state. The high resolution of the microscope revealed many details not seen before, but there was no other method to check if there were artefacts. Only a kind of interdisciplinary research could give a fair assurance that the interpretation could be valid. Several sources of specimen modifications could be introduced during fixation, dehydration, embedding or during observation under the electron beam. In order to get some indications on these questions I decided to do some experiments with tissues which were dehydrated after freezing. Together with my assistant H.R. Müller, we built a special apparatus to embed the freeze-dried tissues in plastic. The tissues were frozen in liquid propane and subsequently dehydrated by sublimation in a glass vessel kept at a temperature of -40° C to -70° C under vacuum (22). The dried specimen was embedded in a monomeric mixture of methyl- and butyl-methacrylate and polymerization was achieved with ultraviolet light in a cooled air stream. The results were very variable and most of the specimens were destroyed by ice crystals. Basically we could find that the same structural details could be seen as in chemically fixed specimens. While we were involved in these studies, from which we learned a lot about the numerous effects involved in freezing and drying, Steere published a paper in 1957 (23), in which virus crystals were made visible by a freezing replica method. He did the freezing in a cold chamber on a brass block and cleaved the specimen with a scalpel under a dissecting microscope and subsequently a metal film was evaporated onto this plane. In contrast to our method, in which the whole specimen was dehydrated before embedding, which took several days, Steere's technique had the advantage that only the cut surface had to be freed of ice, which could be done very rapidly. The concept of setting up a microtome in a vacuum container seemed to offer a way of producing thin sections from frozen tissues for chemical tests, as well as replicas from the etched surface of the cut tissue block. In 1958 we decided to start on this new project with the aim of producing thin sections and replicas of frozen tissues under vacuum (see chapter by H. Moor).

In looking back to the beginning of electron microscopy in Switzerland, the statement that the first ten years after its introduction were the most fascinating ones seems to be justified. During the period between 1942 and 1952 all the basic techniques for application were achieved. This comprises various methods of metal evaporation, surface replication, fixation, embedding and sectioning of cells and tissues, as well as the necessary improvements of all the instruments which were developed for this purpose (microtomes, vacuum evaporators etc.). Compared to our time, when thousands of scientists are engaged in the use of electron microscopes, only very few people were then participating in these developments. Due to lucky chances and good fortune, I was always in the right place to take part in the early developments of these various methods and their applications in biology. Today, when electron mciroscopy has become a routine technique in several fields of science, the difficulties which had to be overcome in the development of the preparation methods during the early days are not known any more. It is no longer realized how much imagination and creativity, combined with craftsmanship, was necessary to work out a reliable method which could be used for different projects by everybody. At the beginning, only very rarely all the failures were subsequently crowned with a success. Therefore we were in a permanent uncertainty if our efforts were worth the troubles we went through. The success of modern electron microscopy proves that these efforts were not in vain.

References

1. H. Ruska, Naturwiss. **27**, 287 (1939).
2. W. Trenktrog, Thesis, Kiel (1923).
3. E. v.Angerer, Techn. Kunstgriffe bei physikal. Untersuchungen, Vieweg, Braunschweig (1924).
4. F. Krause, Z. Phys. **102**, 417 (1936).
5 M. v.Ardenne, Elektronen Uebermikroskopie, Springer, Berlin (1940).
6. K. Mühlethaler and R. Braun, Ber.Schweiz.Bot.Ges. **56**, 360 (1946).
7. W. Wergin, Koll.-Z. **98**, 131 (1942).
8. A. Frey-Wyssling, Koll.-Z. **100**, 304 (1942).
9. K. Wuhrmann, A. Heuberger and K. Mühlethaler, Experientia **2**, 105 (1946).
10. K.H.Meyer and H. Mark, Der Aufbau der hochpolymeren organischen Naturstoffe. Akad.Verlagshaus Leipzig (1930).
11. K.Mühlethaler, Makromol. Chemie **2**, 143 (1948).
12. K.Mühlethaler, Biochim.Biophys.Acta **3**, 527 (1949).
13. H. Ruska, B. v.Borries and E. Ruska, Arch.f.Virusforsch. **1**, 155 (1939).
14. R.C. Williams and R.W.G. Wyckoff, J.Appl.Phys. **15**, 712 (1944).
15. K. Mühlethaler, Biochim. Biophys. Acta **3**, 15 (1949).
16. D.C. Pease and R.F. Baker, Proc.Soc.Exp.Biol.Med. **67**, 470 (1948).
17. L.H. Bretschneider, Kon.Ned.Akad.Wetenschappen **52**, 654 (1949).
18. D. Danon and E. Kellenberger, Archives des Sciences **3**, 160 (1950).
19. S.B. Newman, E. Borysko and M. Swerdlow, Science **110**, 66 (1949).
20. H. Latta and J.F. Hartmann, Proc.Soc.Exp.Biol. N.Y. **74**, 436 (1950).
21. E.F. Fullam and A.E. Gessler, Rev.Sci.Instruments **17**, 23 (1946).
22 H.R. Müller, J. Ultrastruct. Res. **1**, 109 (1957).
23. R.L. Steere, J. Biophys. Biochem. Cytol. **3**, 45 (1957).

History of Electron Microscopy
in Switzerland
Edited by John R. Günter
© 1990 Birkhäuser Verlag Basel

CHEMICAL ELECTRON MICROSCOPY IN BERNE

John R. Günter
Institute for Inorganic Chemistry
University of Zürich

The professors of chemistry at the University of Berne have been interested in the methods of electron microscopy at a very early date, as their school had a long standing tradition in the chemistry of solids, and in particular of finely dispersed and colloidal systems.

The main exponent of these interests was Professor Walter Feitknecht, director of the Institute for Inorganic, Analytical and Physical Chemistry, supported by Professor Kurt Huber (Physical Chemistry) and Professor Rudolf Signer (director of the Institute for Organic Chemistry). A document of their early interest are two reports dating from 1942 and 1943 signed by Ernst Ruska on investigations of nickel hydroxide and vanadium pentoxide colloids performed at the "Laboratorium für Uebermikroskopie,

Siemens & Halske A.G., Berlin" (1). The results on $Ni(OH)_2$ are also included in an early publication (2). Further support for the plans to install an electron microscope in Berne came from the physicist F.G.Houtermans and the zoologist F.E.Lehmann (see also article by E.Kellenberger on the "Bernese connection"). Already in 1945, W.Feitknecht published a review on the investigation of the structure of colloidal substances by means of X-rays and the electron microscope (3), the illustrations of which also come from the Siemens laboratory. Early contacts existed also between the Bernese chemists and the staff of the Trüb, Täuber & Cie. AG in Zürich, as is documented by a publication co-authored by L.Wegmann (TTC) and H.Studer (Univ. of Berne) (4).

A laboratory for electron microscopy was indeed set up already in 1946 in the institute of W.Feitknecht, and one of the first electron microscopes from Trüb, Täuber & Cie. AG was installed in the same year. This laboratory also owned a TTC electron diffractograph (a prototype of the Finch type, equipped for electron diffraction at grazing incidence). In charge of this laboratory was H.Studer, who also cooperated with Prof. A.Frey-Wyssling (see chapter by K.Mühlethaler) and other biologists from different universities. When H.Studer left Berne for the USA around 1955, F.Rüfenacht became his successor for a short period. For the next few years (about 1955-1960), M.Brönnimann became responsible for the chemical electron microscopy. He was succeeded until 1966 by H.R.Oswald, who became director of the Institute for Inorganic Chemistry at the University of Zürich in this year. Since then, the Bernese laboratory is directed by R.Giovanoli, the first Ph.D. student of H.R.Oswald, and he continues the solid state chemical applications of electron microscopy. In Zürich, Oswald also built up a laboratory for chemical electron microscopy, together with his student J.R.Günter, and the Bernese tradition is thus also continued in this university.

The early electron microscopical work in Berne is documented by a number of publications by the three above mentioned professors of chemistry and their students, mainly about colloids of hydroxides and hydroxide-salts (e.g.(5)) (group of W.Feitknecht), Cellulose (6), textile fibres (7) and macromolecules (8), including the use of metal shadowing and two stage replicas (group of R.Signer), and of thin metal oxide layers (e.g.(9)) (group of K.Huber).

Of methodological importance is a short but fundamental paper by H.Zbinden (from his Ph.D. thesis in the group of K.Huber in 1947)(10), in which it is pointed out for the first time how important the electrical surface charge of the supporting films for electron microscopic specimens is, and how it can be influenced.

References

1. R.Giovanoli, personal communication.
2. W.Feitknecht, R.Signer and A.Berger, Kolloid-Z. **101**, 12 (1942).
3. W.Feitknecht, Vierteljahresschrift d. Naturforsch. Ges. Zürich **90**, 161 (1945).
4. L.Wegmann and H.Studer, Chimia **4**, 26 (1950).
5. W.Feitknecht and K.Maget, Z.anorgan.Chemie 258, 151 (1949).
6. R.Signer, A.Aeby, F.Opderbeck and H.Studer, Monatsh.f.Chemie 81, **232** (1950).
7. R.Signer, H.Pfister and H.Studer, Makromolek. Chemie **6**, 15 (1951).
8. R.Signer, Bull.Soc.Chim.France **1949**, 663.
9. K.Huber and B.Bieri, Helv. phys. Acta **21**, 375 (1948).
10. H.Zbinden and K.Huber, Experientia **3**, 452 (1947).

History of Electron Microscopy
in Switzerland
Edited by John R. Günter
© 1990 Birkhäuser Verlag Basel

EARLY TIMES OF ELECTRON MICROSCOPY IN GENEVA
(1944 to 1964)

Eduard Kellenberger
Biozentrum, University
Basel

I. 1944-1952: The Struggle with Instruments

The early electron microscopy in Geneva starts with Jean Weigle*, Professor of Physics. He had been before at the University of Pittsburgh and was what one calls today a solid-state physicist. He was then using X-ray diffraction (4) and electron microscopy, particularly of crystals (20, 21). He started with Paul Dinichert** the study of electrostatic lenses by using the rheographic trough. This must have been around 1943-1944, when Paul Dinichert had obtained a fellowship of the Swiss Medical Academy. At this time a collaboration with Trüb, Täuber & Co. (TTC) in Zürich is likely to have started as well. A written convention must have existed but neither Paul Dinichert, nor Werner Villiger in the Archives of TTC, nor myself could find any written documents as yet. Paul Dinichert found the following in a report of the Geneva commission of electron microscopy (15.1.46), which was composed of the professors J.Weigle, D.Wyss, F.Leuthard, E.Guyénot and E.Bujard:

> "Elle (la commission) se propose d'utiliser pour des recherches biologiques le microscope électronique <u>déposé</u> à Genève par la maison Trüb, Täuber & Co. à Zürich. Pour ces recherches la commission s'est assuré de la collaboration de M.Paul Dinichert, docteur en physique, qui s'est spécialisé dans l'emploi du microscope électronique."

* Swiss citizen, born in Geneva 1901, died in Pasadena, California, 1968. Ph.D. in Geneva in 1923. Research in Pittsburgh, USA, from 1931-1948 Professor and Director of the Physics Institute in Geneva. Research with X-ray diffraction analysis. Studies on the dynamic theory of X-ray diffraction (1968). From 1949 until his death, research associate at the California Institute of Technology, Dept. of Biology with M.Delbrück, where he did research on bacteriophages which later led the ground for molecular biology. Summer months in Geneva, where he put his efforts in building up the molecular biology in Switzerland together with his former student, E.Kellenberger.

** Swiss citizen, born in Berne 1914. Studies of physics in Berlin, Ph.D. work with P.Debye (1937-1939) on paramagnetism at very low temperatures. Work interrupted by the war. Continued in Geneva with J.Weigle on crystalline transformations investigated with X-rays (1940-1942) and later by measuring dielectric constants (1942-1944). Electron microscopy is described in the text. 1947 appointed as physicist to the Laboratoire Suisse de Recherches Horlogères (LSRH); 1957 chef de recherche and since 1962 director of this institution. Retired 1980.

These sentences clearly show that J.Weigle had already at that time seen the particular importance of electron microscopy in its applications to biological problems.

When the electron microscope arrived end of 1946 in the Institute of Physics, it was assembled by Pierre Denis with the help of other members of the institute. Paul Dinichert was then in charge of it. A little later, also an electron diffractograph was delivered for which I was foreseen to be responsible. Things moved rapidly at those times and instead of me, it was J.Hoerni (later in the USA) who was charged while I became assistant of Paul Dinichert for electron microscopy.

At those early days, experiments were made mainly in view of ameliorating the electron microscope. Nevertheless, applications were not neglected. The watch industry had some problems of corrosion on the dials of watches and others with the bore-holes of rubis.

Jean Weigle had contacts with Zworykin (50) and later also with Wyckoff (48) who visited us and left us some preparations of tobacco mosaic virus (TMV) for resolution tests. The collaboration of Paul Dinichert with biologists started with Mark Zalokar, then associate of Emil Guyénot, the famous zoologist of Geneva (14).

In 1947 Paul Dinichert left for Neuchâtel, where he made his career in the "Laboratoires Suisses de Recherches Horlogères" and at the University. In 1962 he became director of this Institution. Electron microscopy was no longer among his duties.

When Dinichert left, I was put in charge of the Genevan electron microscopy. Under the demanding guidance of Jean Weigle I made -until 1950- many experiments for improving the electron microscope itself. This was done in close collaboration with Giovanni Induni, the inventive, extremely original chief engineer of Trüb, Täuber & Co. of Zürich. Our main concern was the correct alignment of the beam and of the three lenses of the microscope. The electron gun was constituted of Induni's cold cathode (see chapter by W.Villiger). The prototype of Geneva had two electrostatic lenses and therefore could not vary the magnification (about 12000 times). We therefore constructed a high voltage divider so as to be able to work also at a lower magnification. This was particularly important for one of our major collaborations that started about 1947-48 with E.Rutishauser, professor of pathology, and his numerous students and collaborators about whom we will come to speak later again.

The improvements made on the electron microscope were not published, but are the topic of numerous reports to TTC. Our focussing system, based on two or three holes in the condensor aperture gave double or triple images out of focus. TTC took a patent for it, which prevented later Siemens to introduce this system (rediscovered by them) in their microscopes.

The technical developments on the instrument were crowned by the design of a new electrostatic objective lens. The resilience of various insulating materials against breakthroughs at high voltage was studied, and the geometry of the electrodes reconsidered. The important limit of resolution through astigmatism had just become known for the electrostatic lens (7) and we immediately saw the advantages - in this respect - of the electrostatic over the electromagnetic lens. While the astigmatism of the first depends on the mechanical precision only, for the latter the heterogeneity of the magnetic permeability of the iron of the pole piece had to be considered in addition. At the early time stigmators were not yet available, and the electrostatic lens had thus a better chance to be produced with as little astigmatism as possible. We knew at that time already that the astigmatism of our lenses was not only due to a lack of mechanical precision but also the consequence of electrical discharges at the edges of the central hole in the middle electrode, possibly indirectly by the resulting electro-erosion that occurred at some regions on the edges of the bore. We found collaboration with the SIP (Société Genèvoise d´Instruments de Physique, the producer of the standard meter and of the world famous coordinate drill press). With its chief mechanic, M. Schmid, we elaborated a way to produce a middle electrode the central hole of which had less of 0.05 µm ellipticity. It was made with the diamond in one single positioning on the lathe. This lens proved to be excellent; it is still in the electron microscope which is now at the "Musée de la Science" in Geneva. TTC however found the price asked by SIP too high, and made their lenses themselves; they had a sharp edge, which became electro-eroded after a few hours of use and thus produced enormous astigmatism. The lens had then to be taken out to be repolished, by which substantial ellipticity was introduced. All that made life very hard and we even had to work night-shifts. Frequently, I put the microscope in working condition overnight, such that my colleagues of the early morning shift could use it again for the biological applications.

In 1950, Jean Weigle stopped the formal collaboration with TTC and put all emphasis on the biological applications which, as said above, had started already before. Kellenberger, nevertheless, maintained very friendly contacts with TTC, particularly with Lienhard Wegmann who was later in charge of its electronic and vacuum-related instruments (see chapter by W.Villiger). On the basis of suggestions from Geneva, a new prototype was delivered (1951/52) to Geneva at half the price. It had a double electrostatic objective lens and a magnetic projector lens, which allowed continuous variation of the magnifications. Lower magnifications were accompanied by such heavy distortions of the image that the micrographs were useless for publications. This prototype was a precursor of the model KM4 of the TTC microscopes (see Villiger).

As formerly mentioned, several biological projects had started before 1950 already. One of them concerned an intensive collaboration with E.Guyénot on the quiescent chromosomes of the ovocyte of triton (Fig.1).

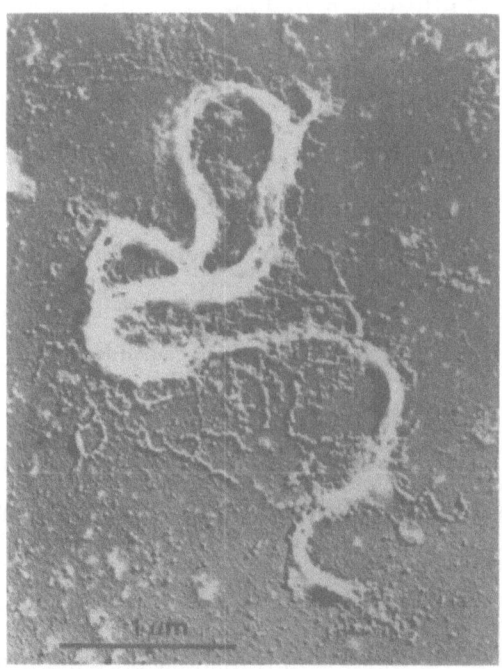

Fig.1:   Quiescent ovocyte nucleus: isolated chromosomes by microcentrifugation. Geneva, 1952. (13, 16)(Galland and Kellenberger).

It was carried out mainly by his graduate student, Mathilde Galland (13, 16, 17). I contributed to this work by inventing a new specimen preparation method: the microcentrifugation (22). It was very much later reinvented by O.Miller; it allowed this author to make his fundamental discoveries of the modes of transcription in prokaryotes and eukaryotes (42). Already at that time I was very much interested in genetics and chromosomes, but E.Guyénot warned my thesis advisor (J.Weigle) and myself that biology was too difficult for a physicist. Jean Weigle had at that time discovered the fundamental and pioneering work of Max Delbrück and his followers (26), which fundamentally influenced the scientific career of both of us. Jean Weigle had already visited Delbrück at Caltech in Pasadena and reported their new findings to my own small group in Geneva. We looked at bacteriophages which we produced ourselves (with Grete Kellenberger-Gujer). Together with Jean Weigle we decided that eventually we should study the intracellular development of bacteriophages by electron microscopy, but so little was known of the anatomy of bacteria that we felt urged to first know more about the uninfected cells. This led to very early collaborations with G.H.Werner (later research director ath Rhône-Poulenc, France) and Valentin Bonifas (later professor of medical microbiology in Lausanne). With the help of the foundation Fritz Hoffmann-La Roche for encouraging team work and in collaboration with F.Chodat, professor of botany, we were able to continue and progress in this line of research (23). My technician, Andrea de Stoutz, and I made shifts day and night to follow the bacterial cultures made in the top floor of the Instiute of Botany, in order to prepare and observe samples in the ground floor of the main building of the university (rue de Candolle) where the electron microscope was located. In the darkness of the night we both were frequently frightened by the busts of the famous late professors placed along the staircase!

We were able to invite two among the scientists that contributed most to the knowledge of the genetic apparatus of the bacteria, the nucleoids, to give us lectures and advice: Berthe Delaporte from France and Gerhard Piekarski from Germany. Both had been pioneering in this field by developping staining methods for light microscopy. At a symposium in Rome (1953) and another in Cambridge, I met also Carl Robinow and R.G.E.Murray (for references see (29)), who later spent sabbatical periods in our laboratory and who had been most influential in my career.

For other research projects (fine structure of teeth: Prof. Held; of bone: Prof. Rutishauser; and for metallurgical purposes) I had developped a particular method of replication (reviewed in (37)) (Fig.2).

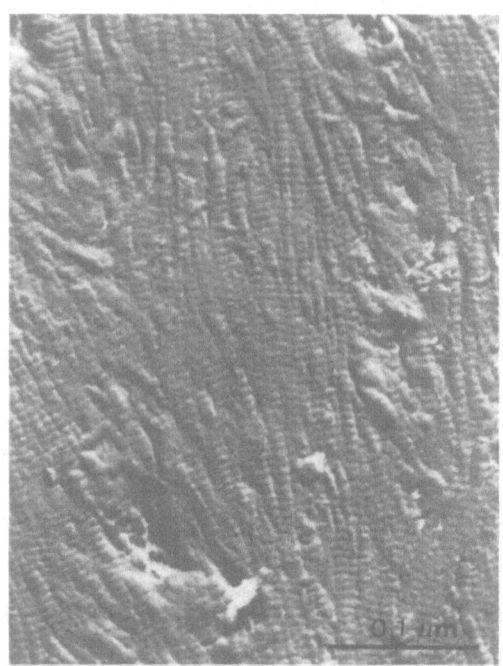

Fig.2:   Structure of bone, by the celluloid replica technique. Geneva, 1952. (33, 37, 44) (Rouiller, Huber and Kellenberger).

The investigation of surfaces by means of replicas was current at that time. Later, SEM was replacing this approach more or less entirely. The main biological application became concentrated on bone. E.Rutishauser encouraged several of his collaborators to engage in electron microscopy with me; they all learned to use the electron microscope and to prepare the specimens, although Lucie Huber (dipl.phys.) of my group helped them as a specialist of replication and microscopy (44).

Some professors and their collaborators followed my teaching of electron microscopy. These early colleagues of the medical faculty, later on, all became famous scientists. Charles Rouiller (1922-1973), later professor of histology in Geneva, Guido Majno (professor at the department of pathology, University of Massachusetts, USA), Pierre Vassalli, later professor of pathology in Geneva. Their major contributions will be considered elsewhere in this book (chapter by Y.Kapanci).

A graduate student in medicine of Prof. Rutishauser, David Danon from Israel, was also delegated to my laboratory. He constructed with me and Jaques Bron (a highly gifted mechanic in the Physics Institute) one of the first microtomes designed ad hoc for electron microscopy (12) and which was commercialized for some time by TTC. David Danon developped new wax mixtures for embedding and applied them to various biological problems (see (51)). After having obtained the price for the best thesis of the year (11), he returned to Israel, where he installed the first electron microscope at the Weizmann Institute. He had numerous students and is considered as the "father of Israeli electron microscopy". David Danon on one side and myself together with J.Bron on the other, both independently improved the microtome so as to adapt it to the use with modern resins (25). His was commercialized, but for ours we renounced, sinde TTC (who had produced the first model) opted for such a small market only that we withdrew from the project which would have been subsidized by a federal fund for industrial research (Fonds Zipfel). One has to be aware that at that time microtomes for electron microscopy just were appearing; the only other microtome with mechanical advance (43) had just been presented and was to be produced by Sorvall.

Several students in physics were able to make their practical diploma (then "licence") work with me. Only one of them, Alain Gautier, stayed with electron microscopy. He was appointed head of the newly created "Centre de Microscopie Electronique" of Lausanne (1955), after having passed a period with W.Bernhard in Villejuif, Paris (see also chapter by E.Kellenberger on the "Bernese Connection").

I had also a close collaboration with Kurt H.Meyer, professor of organic chemistry, who, during the war, had been called from Germany to Geneva. He had been director at Bayer Leverkusen. He was a world renowned specialist of natural and artificial polymers (40), very interested in cellulose and studied with me the only one of animal origin, the tunicine (41). In the field of celluloses he was in a friendly competition with A.Frey-Wyssling in Zürich (see chapter by K.Mühlethaler). They had different working hypotheses on the way the cellulose molecules formed biological structures. Electron microscopy settled the question by proving that the molecules formed fibers which were not anastomosed.

The beginning of the fifties approached when the Institute of Physics entered new premises in a new building. At that time, Jean Weigle decided to renounce to his professorship in Geneva and to accept the position of a research associate at Caltech with Max Delbrück. His contract allowed him to spend four months every year in Europe. He was most of this time in my laboratory in Geneva, although he maintained close contacts with the Paris group of Lwoff, Jacob and Monod. He invited many colleagues of the phage group for visits and collaborative work to Geneva (Siminovitch, Stent, Bertani, Weidel, Dulbecco). My wife Grete and I were invited for a few weeks to work in Paris in order to make some electron microscopy studies with the lysogenic phage system of F.Jacob. We had just before invented a new specimen preparation method for observing the *in situ* lysis of phage producing bacteria (31) (Fig.3). While we prepared the specimens at Pasteur, the electron microscopy was done with the TTC-microscope of W.Bernhard in the Cancer Institute in Villejuif, headed then by Ch.Oberling (see chapter on the "Bernese Connection").

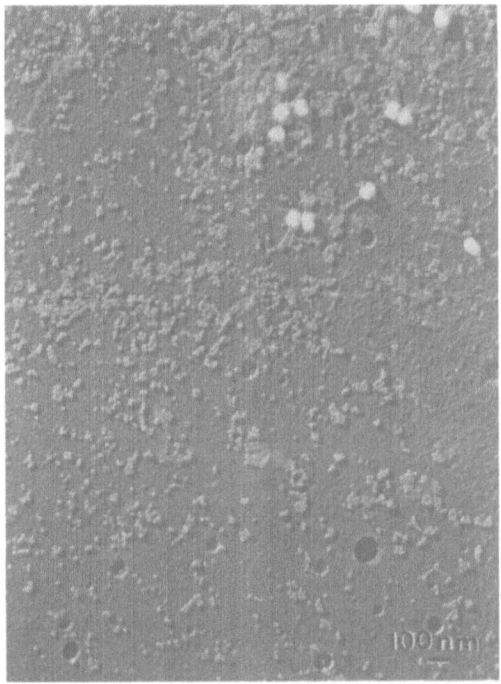

Fig.3:   Part of an *in situ* lysis of a T4-infected *Escherichia coli*.
          With polysomes and phage DNA pool. Geneva, 1955. (G. and E. Kellenberger).

Very little of the work of this time was published, because too many completely new features needed a backing by physiology, genetics and biochemistry, which came only later. As a physicist, I was not satisfied with pure speculations and preferred to await further clarifications. (For these aspects, compare Kellenberger (26) and later publications of the Geneva group (27, 30)).

In the early fifties I was given the opportunity by Jean Weigle and Kurt H.Meyer to consider buying a new electron microscope and I could visit several European groups to find out about the most adequate commercial brand. We knew already at that time that the biological specimens were easily altered by beam-induced damage. We had seen such alterations on the tunicine. I thus took such a specimen along with me and visited laboratories in Paris, Delft and Düsseldorf. These visits initiated long-lasting friendships even when we finally decided in favour of the RCA EMU 2, which clearly produced the least damage, because its designer, the physicist James Hillier (later research director of RCA) had been fully aware of biological specimens (19). He had limited the intensity of the beam by the condensor aperture, acting before the specimen, while in most of the European electron microscopes the objective aperture - in the centre of the lens - was still limiting the intensity besides correcting the spherical aberration.

The new microscope was delivered around 1951/52 into the new building of the Physics Institute. We could participate in the interior design of this building and could thus prepare a very suitable laboratory for electron microscopy and for what was later to become molecular biology. The latter was always very strongly supported and aided by Jean Weigle with his regular returns into my laboratory.

With the new microscope I could finally prepare the good quality electron micrographs needed for publications. This was successfully done at a symposium on the bacterial nucleoid in Rome (24); this paper was also the basis of my thesis.

With the electrostatic lenses of the TTC-microscopes and with our first type of microtome we had very little success with resin sections. I was therefore given the possibility to visit Fritiof Sjöstrand in Stockholm, who was then the leader in thin sectioning. During several weeks I learnt honing razor blades. Back home it became rapidly clear that the electrostatic lens with its high chromatic error was not suitable for thin sections. With the RCA EMU 2, this matter was immediately improved and a period, concentrated on sectioning, arose, which will be discussed in another chapter (cytology) and in the next section. It allowed for a renewed collaboration with my friend Charles Rouiller who, after several postdoc-years in Paris, had come back as professor of histology. For many years he used our RCA microscope for his studies on the liver (see chapter by Y.Kapanci). He got his own electron microscope only later. In the meantime he shared also the new Siemens electron microscope with us, which we got from the Swiss National Science Foundation at the end of the fifties. For the biological applications at that time, a comprehensive review has been written by Kellenberger and Rouiller (32).

Before we enter the second period, a few lines are perhaps justified to evoke the fate of metallurgical applications. After the few collaborations with the watch industry (mentioned above), we had an excellent collaboration with Dr. Saulnier of Péchiney in Chambéry (France). He had bought a TTC microscope in the late 40´s and had introduced our new method of replication. With this outfit he participated with his company in establishing the Be-Al alloys which were used later as nonmagnetic springs in our watch industry! I therefore felt a strong urge to push also our Swiss metallurgy into the use of electron microscopy. I had a Swiss engineer at hand who had a background in metallurgy and who worked some time with us. With him I wanted to establish a small laboratory for keeping abreast also in this new field of applications of electron microscopy. I had no success in finding the necessary money and gave up. In another chapter of this book (by W.Bollmann), the later raise of metallurgical applications will be described.

II. The Participation of the Geneva Electron Microscopy in the Rising Field of Molecular Biology (1952-1964)

In this section we now report about the early, first applications of electron microscopy to the study of bacteriophages, their structure and their intracellular development. The instruments were no longer the center of our activities, although with Jacques Bron we still made a fair number of constructions: The two TTC prototypes ("Arthur" from 1946 and "Marianne" from 1952) were melted into "Babar", which had many modern features and facilities for work. This microscope was still in constant use, until much later, when it was given to the engineering school of Geneva. Still later, when they got the RCA ("Johnny"), Babar went to the Science Museum of Geneva, where it still stands.

In the new premises we also built an evaporation unit according to most modern vacuum techniques. We had to import the high power oil diffusion pump directly from the USA, because most European manufacturers were far behind in vacuum technology. This unit was also suitable for the William´s technique of freeze-drying (47).

It might be recalled here that TTC had based all their equipment on Induni´s very interesting version of the molecular pump of the Holweck type. We had also constructed our very first evaporation unit with this pumping system; later a diploma student of mine had - in the previous period - already built a new unit with an American oil diffusion pump.

At the very beginning of the new period, the Rockefeller Foundation paid us a preparative Sorvall ultracentrifuge, as well as a prolonged visit of Thomas F. Anderson to our laboratory with the purpose of introducing his new critical point method to us (2, 3).

Our group consisted now of two diploma physicists, Lucie Huber and Janine Séchaud and my first wife Grete on a position as technician, but with much higher abilities. After having first helped me in electron microscopy, she very soon became the main collaborator of Jean Weigle in molecular genetics. She was therefore a figure of prime importance in the common successes of Jean Weigle and me in introducing in Switzerland what was later to become molecular biology. The first group was soon joined by Antoinette Ryter (a dipl. chem.) and Werner Arber (Dipl.Nat.ETH), both being paid by my first research grant, which I obtained from the Swiss National Science Foundation.

Werner Arber was recommended to us by the late Paul Scherrer, professor of physics at the ETH, with whom I also had made the first part of my physics studies. Werner was in charge of "Arthur", while I used "Johnny". Together we discovered the tail sheath contraction of phage T_{even} *in vitro* after oxidative treatment and *in vivo* on empty cell envelopes (33) (Figs. 4 &

102

5). The paper still contains micrographs made with both microscopes. Later, Werner Arber (Nobel Prize in 1978 for his pioneering and assiduous work on restriction enzymes which are now the basis of the *in vitro* recombination on DNA) worked on the physiology of transducing phages which was his thesis project (Fig. 7). In this frame he still used electron microscopy in one of his many papers (5).

At this time I had also a very successful collaboration with Wolfhard Weidel, a German pioneer of phage work in Tübingen (Germany). We demonstrated the bacterial receptor of bacteriophage T5 (46). For the

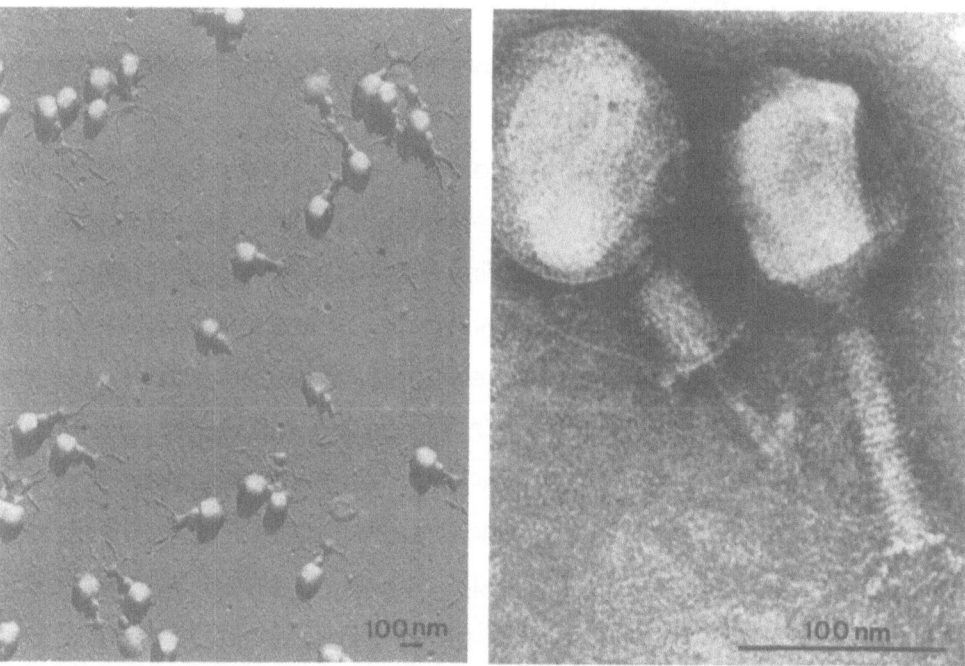

Fig. 4:   Oxidized T$_{even}$ bacteriophage. Geneva, 1955, arch.nr.16488. (E.Kellenberger and W.Arber).

Fig. 5:   Bacteriphage T$_2$, a normal one together with one with contracted tail sheath. The contraction is a normal event in the process of infection. Geneva, 1964, arch.nr.31915. (E.Boy de la Tour and E.Kellenberger).

exact determination of its shape we used the freeze-drying technique of Williams and Bachus (Fig. 6).

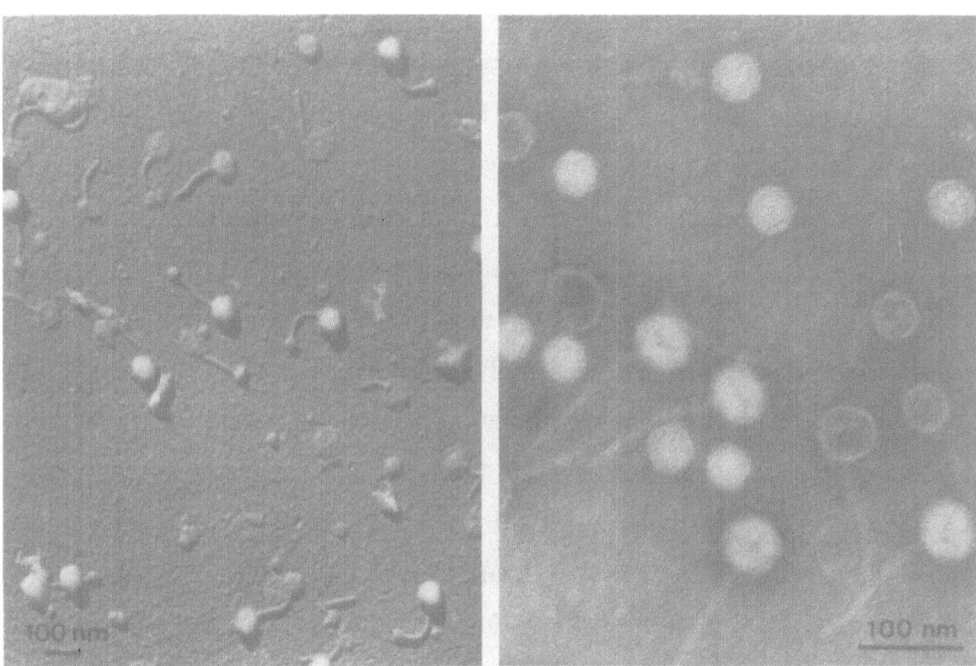

Fig. 6: Phage T5 with its receptor. Geneva, 1954. Arch.nr.15046. (E.Kellenberger and W.Weidel).
Fig. 7: Bacteriophage lambda with its small variant, p-lambda, full and empty. Geneva, 1962. Arch.nr.26506. (E.Kellenberger).

In the meantime K.H.Meyer had tragically died, but we had maintained an interest in polysaccharides. Collaborating with L.Laszt from Fribourg, Werner Arber and I investigated liver glycogen by using also the new technique of freeze-drying. We were able to establish that glycogen was not cigar-shaped, as often claimed in the literature, but rather a sponge-like, highly hydrated globular particle (6).

With Antoinette Ryter, I focussed on improving the techniques of thin sectioning bacteria with and without intracellularly developing bacteriophages. We introduced new embedding resins (see corresponding chapter in this book) and in particular established conditions for the OsO_4 fixation, by which DNA containing plasms could be adequately fixed (ref. in (34)). The fixation was a prerequisite for studying non-infected and infected bacteria. The fixation procedure was used for a long time by bacteriologists and once even brought us the (dubious!) glory of a citation classic (28).

Janine Séchaud and Grete Kellenberger-Gujer joined now in the study of intracellular phage, which was the start of a long series of works which is summarized elsewhere (8, 27, 30). Through the success of these papers, the Geneva laboratory became now very attractive for American and other fellows, who filled our premises since the early fifties. It is not the place here to enumerate all the names of postdocs and sabbatical professors; only a few of them, who were closely concerned with electron microscopy should be recalled: Charly Brinton, with his fundamental work on the pili (fimbriae) of bacteria, of which only little is published (for references see (9)), although he still continues this work making important investigations as a professor in Pittsburgh. A very intimate detail of his experiences in Geneva might be of interest to the reader: He used "Arthur" for his studies and one late evening he arrested the vacuum, but forgot to switch off the high voltage! The lenses were not destroyed, but the voltage supply was dead. After having previously been mounted in air, we had just put it all together into a newly built big tank of oil, and now poor Charly had to dive deep into the oil to repair and replace parts!

Another postdoc, Naomie Franklin, who, although not being primarily a microscopist, performed electron microscopy by herself. She did that at a particularly bad time, when the service engineer - against Janine Séchaud's and my opinion - made a wrong diagnosis of failures in the RCA microscope. He believed that the lack of sharpness was due to external fields and we had to move the microscope through all the possible sites in the basement of the building. More than a year later a new service engineer discovered a burnt self-induction coil in the high voltage circuit, confirming our diagnosis of a failure in the electronic part of the instrument. Naomie, nevertheless, made important determinations on the functions of the tail fibers of bacteriphage T4 by counting painstakingly the numbers of fibers extended (38). Another future Nobel Prize winner co-authored this paper by participating very closely in the design of the experiments: Nils K.Jerne (Nobel Prize in 1984 for his pioneering work on the selection theory of immunology), then an employee of WHO, who was also sort of a permanent visiting professor of our Department (Söderquist, Biography of N.K.Jerne, in preparation).

After emphasizing the importance of foreign visitors for the Geneva laboratory, we should not forget to mention that the intensive collaboration with the Genevean biologists continued. With Gilbert Turian (member of the Institute of Botany and later professor of microbiology), we studied the paranuclear body of Allomyces (45). We had then just discovered what we call today the aggregation-sensitive chromatins in bacteria and in the DNA pool of vegetative phage (36) and were interested to extend the study by comparing with other chromatins. With Benigna Blondel (later research associate in histology), we studied synchronized cultures of human cells for understanding more about the metabolically active DNA of the interphase when compared to condensed chromosomes (8a). In this frame was also a larger research project with Gérard de Haller (Institute of Zoology, now professor in that discipline) on the crazy chromosomes of Dinoflagellates (18). This line of research intensified also our collaborative work with my friend Charles Rouiller, who, as a professor of histology, had obviously also more important tasks (his activities are discussed in the chapter by Y.Kapanci).

Of this period, we should also mention our attempts in the study of the crystallographic arrangement of the protein subunits on the capsid of bacteriophage. We made use of a tubular variant of this capsid which allowed to introduce crystallographic methods into biology. We discovered that the two layers of the completely flattened tube produced a moiré pattern, by which crystallographic parameters could be determined (35). This discovery made me earn the honorary membership of the French EM Society, but also a friendship with Aaron Klug, which was associated with a very fierce competition! He introduced the optical diffraction which he applied most successfully to our material (39, 49). We attempted three times to introduce and establish these methods also in Switzerland, but were successful only the third time, when in 1985 my former students, respectively postdocs, Uli Aebi and Andreas Engel were called back for heading the privately funded Maurice E.Müller Institute of High Resolution Electron Microscopy (MIH) housed in the Biozentrum in Basel.

For the previous attempts in Geneva, E.Boy de la Tour merits a particularly honorable mention. As a professional photographer, he became an outstanding microscopist and later an expert on the optical diffractometer of which he built two prototypes with me and with the help of the mechanics of the Physics Institute, one being the first in a bent configuration and which was rebuilt twice later in Basel (ref. in (1)) and also elsewhere.

Among the electron microscopists who spent a postdoctoral and/or a sabbatical period in the Geneva laboratory, we should mention Fred Eiserling (US citizen) and Lucien Caro (French citizen), The first continued the electron microscopy of phages at UCLA and became famous for it (15). The second had introduced a well known method of autoradiography (for ref. see (10)) and later was renowned for his work on the molecular biology of DNA. He became my successor in Geneva, when I accepted to move to Basel.

References

1. U.Aebi, P.R.Smith, J.Dubochet, C.Henry and E.Kellenberger, J.Supramol.Struct. **1**, 498 (1973).
2. T.F.Anderson, J.Appl.Phys. **21**, 724 (1950).
3. T.F.Anderson, Trans.NY Acad.Sci. **16,** 242 (1954).
4. W.Arber, in: Actes de la Société Helvétique des Sciences Naturelles, 293 (1969).
5. W.Arber and G.Kellenberger, Virology **5**, 458 (1958).
6. W.Arber, E.Kellenberger and L.Laszt, Kolloid Z. **150**, 123 (1957).
7. F.Bertein, H.Bruck and P.Grivet, Ann.Radioel. **2**, 249 (1947).
8. L.W.Black and M.K.Showe, in: Bacteriophage T4 (C.K.Mathews, E.Kutter, G.Mosig and P.B.Berget, eds.), Amer.Soc.Microbiol., Washington DC, USA, 219 (1983).
8a.B.Blondel, Exp. Cell Research **53**, 348 (1968).
9. C.C.Brinton, Trans. NY Acad Sci. **27**, 1003 (1965).
10. L.Caro, in: Progress in Biophysics and Molecular Biology 16, Pergamon Press, 171 (1966).
11. D.Danon, Thesis, Kündig Geneva (1952).
12. D.Danon and E.Kellenberger, Arch. Sciences **3**, 169 (1950).

108

13. M.Danon, E.Guyénot, E.Kellenberger and J.J.Weigle, Nature
    165, 33 (1950).
14. P.Dinichert, E.Guyénot and M.Zalokar, Rev.Suisse de Zoologie **54**,
    283 (1947).
15. F.A.Eiserling, in: Bacteriophage T4 (C.K.Mathews, E.M.Kutter, G.Mosig
    and P.B.Berget, eds.), Am.Soc.Microbiol., Washington DC, USA, 11
    (1983).
16. E.Guyénot and M.Danon, Rev.Suisse de Zoologie **60**, 1 (1953).
17. E.Guyénot, M.Danon, E.Kellenberger and J.Weigle, Arch. der Julius
    Klaus Stiftung **25**, 47 (1950).
18. G.de Haller, E.Kellenberger and Ch.Rouiller, J.Microscopie **3**, 627
    (1964).
19. J.Hillier, Ann.Rev.Microbiol., 1-20 (1950).
20. J.Hoerni, Nature **164**, 1045 (1949).
21. J.Hoerni and J.Weigle, Nature **164**, 1088 (1949).
22. E.Kellenberger, Experientia **5**, 253 (1949).
23. E.Kellenberger, Z.wiss.Mikr.& mikr.Technik **60**, 408 (1952).
24. E.Kellenberger, in: Symposia Citologia Batterica (Roma), Suppl.
    Rendiconti Istituto Superiore di Sanitá, Roma, 45 (1953).
25. E.Kellenberger, Experientia **12**, 282 (1956).
26. E.Kellenberger, in: Phage and the origins of molecular biology,
    (J.Cairns, G.S.Stent and J.D.Watson, eds.), Cold Spring Harbor
    Laboratory, 116 (1966).
27. E.Kellenberger, BioSystems **12**, 201 (1980).
28. E.Kellenberger, Curr.Contents **7**, 12 (1980).
29. E.Kellenberger, in: The bacterial chromosomes, (M.Riley and K.Drlica,
    eds.), ASM, Washingotn DC, USA, 173 (1990).
30. E.Kellenberger, Eur.J.Biochem. **190**, 233 (1990).
31. G.Kellenberger and E.Kellenberger, Schweiz.Z.allgem.pathol.Bakteriol.
    15, 225 (1952).
32. E.Kellenberger and Ch.Rouiller, in: Médecine et technique (G.Laroche,
    ed.), Albin Michel, Paris, 264 (1954).
33. E.Kellenberger and W.Arber, Z.Naturforschg. **10b**, 698 (1955).
34. E.Kellenberger and A.Ryter, in: Modern developments in electron
    microscopy, (B.Siegel, ed.), Acad. Press, 335 (1964).
35. E.Kellenberger and E.Boy de la Tour, J.Ultrastruct.Res. **13**, 343 (1965).

36. E.Kellenberger, A.Ryter and J.Séchaud, J.Biophys.Biochem.Cytol. **4**, 671 (1958).
37. E.Kellenberger, L.Huber and Ch.Rouiller, Z.wiss.Mikrosk. **63**, 110 (1956).
38. E.Kellenberg, A.Bolle, E.Boy de la Tour, R.H.Epstein, N.C.Franklin, N.K. Jerne, A.Reale-Scafati, J.Séchaud, I.Bendet, D.Goldstein and M.A. Lauffer, Virology **26**, 419 (1965).
39. A.Klug and J.E.Berger, J.Mol.Biol. **10**, 565 (1964).
40. K.H.Meyer, Natural and Synthetic High Polymers, Interscience Publ., NY, USA (1950).
41. K.H.Meyer, L.Huber and E.Kellenberger, Experientia **7**, 216 (1951).
42. O.L.Miller, Sci.Amer. **228**, 27 (1973).
43. K.R.Porter and J.Blum, Anal.Record. **117**, 685 (1953).
44. Ch.Rouiller, L.Huber, E.Kellenberger and E.Rutishauser, Acta anatomica **14**, 9 (1952).
45. G.Turian and E.Kellenberger, Exptl.Cell Res. **11**, 417 (1956).
46. W.Weidel and E.Kellenberger, Biochem.Biophys.Acta **17**, 1 (1955).
47. R.C.Williams, Exptl.Cell Res. **4**, 188 (1953).
48. R.W.G.Wyckoff, Electron Microscopy, Interscience Publ., NY - London, (1949).
49. M.Yanagida, D.J.de Rosier and A.Klug, J.Mol.Biol. **65**, 489 (1972).
50. V.K.Zworykin, G.A.Morton, E.G.Ramberg, J.Hillier and A.W.Vance, Electron Optics and the Electron Microscope, John Wiley & Sons Inc., NY, USA and Chapman & Hall Ltd., London (1945).
51. Comte rendu du premier congrès de microscopie électronique, Paris 1950. Ed.rev.d'optique, 1952.

History of Electron Microscopy
in Switzerland
Edited by John R. Günter
© 1990 Birkhäuser Verlag Basel

THE "BERNESE CONNECTION" OF EARLY PIONEERS IN BIOLOGICAL ELECTRON MICROSCOPY
F.E. Lehmann (Berne), W. Bernhard (Paris) and H. Ris (Wisconsin)

Eduard Kellenberger
Biozentrum, University
Basel

Contradicting the general belief about the Bernese being slow and conservative, they had bought the first TTC electron microscope of a commercial series which was delivered in 1946 to the laboratory of W. Feitknecht, professor of inorganic chemistry (see chapter by J.R.Günter). He found a large echo among his colleagues. With H. Studer, the responsible of the microscope, the new facility had found an excellent pioneer with a broad knowledge and understanding of electron microscopy. He was (as I was too) particularly interested in the methods of specimen preparation and I remember that he was one of the (still) very few who recognised that in the preparation of particles from suspensions their ability to adhere to the surface of the film is of prime importance. It depends on surface charge and hydrophilicity. He thus rendered the collodion film basic (positively charged) by coating it with a thin film of Be by evaporation (19). We took this line up by further (unpublished) work in Geneva, and produced also acidic films by chemical modification of the surface of collodion. These important facts about the adherence of molecules or particles to the supporting film (see (10)) are still largely ignored by the users.

Rudolf Signer (born in 1903, professor for organic chemistry from 1935 to 1972) had very modern views on the nature of biological high polymers and I remember very stimulating discussions with him about the electron microscopy of proteins, particularly of casein.

Probably the first biologist to have started (in 1950) to use the electron microscope in Berne was Fritz E. Lehmann (1902-1970) from the Institute of Zoology.

It is not astonishing then that this breeding ground had stimulated Bernese students as Wilhelm Bernhard (medical student in Berne from 1940-1946) and Hans Ris (diploma of high school teacher from the University of Berne, 1936) to choose electron microscopy as their major tool of research after they had emigrated. These two Bernese pioneers in biological electron microscopy have started electron microscopy only after they had left Switzerland; indeed, the Bernese microscope arrived after their departures.

Fritz E. Lehmann was called in 1929 to Berne as associate of Prof. F. Baltzer, director of the Institute of Zoology. He became private-docent in 1931 and was appointed associate professor in 1940 and full professor in 1948. In 1954 he succeeded F.Baltzer as head of the Institute. Because of a serious illness he retired in 1965 from the directorship and devoted his remaining strength to scientific research. He died in 1970.

His main research was developmental biology (11); he studied in particular the inhibition of cell division by various antimitotic substances. He seems always to have been interested in structural problems. He started therefore to use the Bernese electron microscope shortly before 1950 (12, 13). He attacked immediately the preeminent question about the structure of the constituents of the cytoplasm ("the biological colloids") as had been pioneered just before by optical techniques by Staudinger in Germany and Frey-Wyssling in Switzerland. F.E.Lehmann documented by electron micrographs their postulates, as for instance that the hyaline ground plasma of the cellular cytoplasm must be composed of fibrillar and globular elements.(1, 12, 13). This was, however, just at the same time when Sjöstrand, Palade and Porter had declared that only the fixation with OsO_4 could produce micrographs of acceptable quality. Compared to those of Lehmann´s micrographs, mostly obtained after acid fixations, the dimensions of the OsO_4 fixed fine structures were indeed about 10 times smaller. Not too many people believed in Lehmann´s results (30), except for a relatively large school in Italy which was animated by A. Beirati who had worked together with Lehmann. Why then did we mention these early, contestable results? The main reason for it is the fact that many years later Keith Porter (the same as mentioned above) himself proposed the real existence of what he called the "microtrabeculae" (18), which he considered to be a native fibrous element of the cell which is about 10 times thicker than the microtubuli. This happened after Porter´s name had entered the history by being among the pioneers of intracellular microtubuli (17) long before this important cytoskeletal element had been isolated and chemically characterized.

The microtrabeculae were observed with high voltage electron microscopy in cells which were prepared by the critical point drying. The cells are dehydrated with organic solvents before they are tranferred either into liquid CO_2 or into Freon. At the critical values of temperature and pressure the latter liquid is then transformed into a gas without producing an interface. Surface tension effects can thus be avoided. The aggregation phenomena associated with the substitution of the cellular water by organic solvents are not eliminated. They are the reasons why soluble proteins are aggregated onto fibrous elements of the cytoskeleton and thus causing their thickening. This "artefact" has most clearly been demonstrated by Hans Ris (16, 22), discussed below.

There is no doubt that both Lehmann and Porter had observed aggregation figures or at least artificially thickened, possibly preexisting fibrous structures. In contrast to the structures of Lehmann, the microtrabeculae of Porter had, nearly 40 years later, a large impact onto the scientific community, particularly among the biochemists. The hypothesis that in vivo enzymes might sit on structures and not be in solution as they are in vitro, was and still is considered as an important challenge. I think therefore that it is high time to give also F.E.Lehmann some credit as having been the first who had described, although not named, the microtrabeculae!

Besides the already mentioned book on developmental biology (11) the main contribution of F.E.Lehmann concerns the structure of the nuclear membrane, studied together with Beirati (13a, b).

Lehmann has had very successful students and later collaborators. To name only a very few: Rudolf Weber became his main associate for electron microscopy (27-29). He animated an electron microscopy group in the Zoology Institute which got its own electron microscopes, in 1964 a Hitachi HS-75 and in 1971 a Philips 300. R.Weber became the successor of F.E.Lehmann as director of the Institute of Zoology.

Hans Ris was born on June 15th, 1914 in Berne. He studied high school teacher at the University in Berne. After having reached his diploma in 1936 he emigrated (1938) to the USA where he achieved his Ph.D. in Columbia in 1942. He became an US citizen in 1945. After research and teaching at Yale University, John Hopkins and the Rockefeller Institute, he became professor of Zoology at the University of Wisconsin in Madison in 1946. In the 70's he was, besides K.Porter in Colorado, the second biologist in the USA who was chosen for exploring the possibilities of high voltage electron microscopy in biology.

With Mirski, Mazia, Taylor, Moses, Nebel, Rebhuhn and Prescott, Hans Ris was among the very early pioneers of the structure of chromatin and of chromosomes. He had characterized the "normal" chromatin of higher cells as some 30 nm fibers (20, 21). He had observed very early that, when prepared with less care, these fibers desintegrated into beads. As a critical microscopist, he diregarded these beads as preparation artefacts as they are indeed. As happens frequently, even artefacts can provide new information: as is well known now, these beads turned out to be the well characterized nucleosomes. Micrographs of the "beads on a string" (15) became rapidly famous, although natural, native chromatin is, as Hans Ris showed, a cylindrical fiber. The image of the "beads on a string", as representing a linear sequence of globular nucleosomes, is still preponderant in the mind of most biologists, although in the modern textbooks the models mostly present the correct situation. The merit of Hans Ris, for having shown a correct image of the chromatin, is rarely mentioned or even emphasized. References to his important work both in biology and in preparation methods for electron microscopy are found in the articles (23, 25, 26).

Wilhelm Bernhard was born on November 8th, 1920 in a farm near Worb (Canton Berne). He studied medicine at the University of Berne (1940-46). In 1948 he started his career at the Cancer Institute in Villejuif near Paris. He became a French citizen. He died tragically at an electron microscopy conference in Buenos Aires in 1978.

As a medical doctor, Wilhelm Bernhard was attracted from the beginning by the fundamental problems of genesis and fate of tumor cells. He thus accepted a position at the famous Cancer Institute in Villejuif near Paris. It was headed by Charles Oberling, an Alsatian with very modern ideas. Oberling was one of the first to consider a general relation between virus and cancer (14). As a pioneer, he also saw the potential role of electron microscopy in cell biology. Bernhard thus got at Villejuif a new TTC microscope; the common worries about it brought us together. He also got an RCA EMU 2 later, as we did, which allowed then both of us to enter the international competition. He had discovered very early what he then thought to be filaments in the "ergastoplasm" of cells of the liver and pancreas (5, 6). I remember vividly our argumentation about whether these were real filaments or cross-sections of membrane systems. It was later demonstrated that the ergastoplasm was a membrane system, the endoplasmatic reticulum, studded with ribosomes (2, 6). Together they constitute what A.Claude (later in Belgium) had isolated by centrifugation as microsomes (9). This is all only to say that W.Bernhard was internationally at the very forefront of the rising modern cytology (8).

Geneva had from the beginning excellent scientific and friendly contacts with Villejuif and I remember the story about our prototype of microtome (see chapter by E.Kellenberger on Early Times of EM in Geneva), which Bernhard wanted to try out. We had packed it carefully into one of the wooden boxes which were then used by the students in the painting lessons, and Alain Gautier had to transport it to Paris. In order to pass the customs, it had to be declared as "machine pour détecter le cancer", because nothing similar to an ultramicrotome was known at that time! Later, W.Bernhard, together with Smetana in the CSR and Bush in the USA, were leaders in the cytology of the nuleolus (3, 7, 24).

As discussed in the chapter by A.Gautier, W.Bernhard had constant interests in the cytochemistry on the level of electron microscopy and for the improvement of embedding techniques (see chapter by Kellenberger on Embedding Techniques).

Besides his many internationally well recognized contributions to cytology which became the basis of modern cell biology (see 8, 24), he has also been outstanding by his concerns about the fate and inequities of humanity. These concerns are summarized in a pamphlet which was edited after his unexpected death (4).

References

1. A.Beirati and F.E.Lehmann, Rev.Suisse de Zool 54, 443 (1951).
2. W.Bernhard, Nova Acta Leopoldina 35, 185 (1970).
3. W.Bernhard, in: Cell differentiation in microorganisms, plants and animals (L.Nover & K.Mother, eds.), VEB Gustav-Fischer, Jena, 579 (1977).
4. W.Bernhard, Biol.Cell. 33, XXVII (1978).
5. W.Bernhard, F.Haguenau, A.Gautier and Ch.Oberling, Z.Zellforschg. 37, 281 (1952).
6. W.Bernhard, A.Gautier and Ch.Rouiller, Arch.Anal.Micros. 43, 236 (1954).
7. W.Bernhard, A.Bauer, A.Gropp, F.Haguenau and Ch.Oberling, Exp.Cell Res. 9, 88 (1955).
8. M.Bouteille, J.André, M.Hertzberg, R.Simard and F.Haguenau, Biol. Cell 29, 113 (1980).
9. A.Claude, K.R.Porter and E.G.Pickels, Cancer Res. 7, 421 (1947).
10. J.Dubochet and E.Kellenberger, Microscopica Acta 72, 119 (1972).
11. F.E.Lehmann, Einführung in die physiologische Embryologie, Birkhäuser, Basel (1946).
12. F.E.Lehmann, Rev. Suisse Zool. 53, 141 (1950).
13. F.E.Lehmann, Experiemtia 6, 382 (1950).
13a.F.E.Lehmann, Klinische Wochenschrift 33, 294 (1955).
13b.F.E.Lehmann and V.Mancuso, Exp. Cell Res. 13, 161 (1957).
14. Ch.Oberling, Le Cancer, Gallimard, Paris (1954).
15. A.L.Olins and D.E.Olins, Science 183, 330 (1974).
16. J.Pawley and H.Ris, J.Microsc. 145, 319 (1987).
17. K.R.Porter, in: Principle of biomeolecular organization, (a CIBA symposium), (G.E.W.Wolstenholme and M.O'Connors, eds.), Churchill, London, p. 308 (1966).
18. K.R.Porter and K.L.Anderson, Eur.J.Cell Biol. 29, 83 (1982).
19. E.Ribi and B.G.Ranby, Experientia 6, 27 (1950).
20. H.Ris, in: Regulation of nucleic acid and protein biosynthesis, (V.V.Koningsberger and L.Bosch, eds.), Elsevier Publ.Co., Amsterdam, p. 11 (1967).

21. H.Ris, in: Electron Microscopy (Proc. 9th Int.Congress on EM), (J.M.Sturgess, ed.), Vol.III, 545 (1978).
22. H.Ris, J.Cell Biol. 100, 1474 (1985).
23. S.Sepsenwol, H.Ris and T.M.Roberts, J.Cell Biol. 108, 55 (1989).
24. K.Smetana and H.Busch, in: The cell nucleus, (H.Busch, ed.), Vol. 1, Academic Press, NY, USA, p.73 (1974).
    H.Busch and K.Smetana, The Nucleolus, Acad.Press, NY, USA (1970).
25. A.Szollosi, H.Ris, D.Szollosi and A.Debec, Eur.J.Cell Biol. 40, 100 (1986).
26. H.T.Tsui, K.L.Lankford, H.Ris and W.L.Klein, J.Neurosci. 4, 3002 (1984).
27. R.Weber, Roux's Arch.Entwicklungsmech.Organe 150, 542 (1958).
28. P.Wellauer, T.Wyler and E.Buddecke, Hoppe Seyler's Z.Physiol.Chem. 353, 1043 (1972).
29. P.Wellauer, R.Weber and T.Wyler, J.Ultrastruct.Res. 42, 377 (1973).
30. K.E.Wohlfarth-Botterman, Protoplasma 49, 231 (1958).

III. MATERIALS SCIENCE

History of Electron Microscopy
in Switzerland
Edited by John R. Günter
© 1990 Birkhäuser Verlag Basel

ELECTRON MICROSCOPY AT THE BATTELLE LABORATORIES IN GENEVA

Walter Bollmann
22 Chemin Vert, Pinchat
CH-1234 Vessy, Geneva

Introduction

Since this paper is foreseen to describe contributions to the historical development of electron microscopy, it will contain not only facts but also the ideas and influences which led to the facts. In a scientific paper the results have to be presented as concisely as possible and the theoretical considerations as a strict logical sequence. People outside the scientific community may then get the impression that scientists are all perfect Sherlock Holmeses. In reality the logics is usually the final construction grown out of lots of vague hints, analogies, errors and wrong paths. Once the result is attained in a rather troublesome way, then the work of writing the paper starts which usually takes more time than to find the result. If the paper did exactly describe how the result was obtained, it would (a) become too long and (b) be rather confusing. A scientist is not only paid for finding something but also for explaining to his colleagues what he has found and this in a way that it can be understood with as little effort as possible.

In the following sections I shall write about the direct observations of dislocations by transmission electron microscopy, and applications such as the study of recrystallization of nickel and the phase transformation of cobalt. Further work concerned the study of radiation damage of graphite. Later the work shifted more to the theory although closely related to electron microscopy in the sense that it helped to understand the pictures and to link this understanding with dislocation theory. This field concerns the theory of dislocation networks and the general geometrical theory of intercrystalline boundaries which I shall only briefly mention and give the literature references.

The Development of Transmission Electron Microscopy (TEM)

Since Battelle is a contract research organization, the work was either done on externally sponsored research contracts or on internal projects as a preparation of the field for later external projects. When I started at Battelle in December 1953 my background was a doctorate in nuclear physics and two years work in industry in a high voltage laboratory. My first task at Battelle was to build an electron diffraction equipment (for grazing incidence) for the study of condensed layers of germanium on germanium, which was a project for an electronic industry (Fig.1).

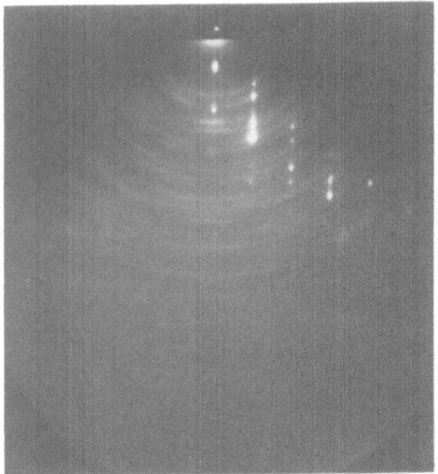 

Fig.1: a) Electron diffraction equipment for grazing incidence. The electron beam moves from left to right. At the left side is the gun with the high voltage equipment (40 kV), to the right the fluorescent screen and a Trüb-Täuber camera for photo plates. The top part is the heatable object holder and the bottom part the evaporation equipment. The vacuum pump acts from behind and in front is an observation window.
   b) Debye-Scherrer diffraction of an essentially polycrystalline deposit.
   c) Kikuchi diffraction of an essentially monocrystalline deposit.

This forced me to learn electron diffraction and to interpret patterns of the Laue, the Debye-Scherrer and the Kikuchi type.

Battelle had then a Philips EM 100 electron microscope which was under the supervision of Dr. J.-C.Courvoisier. In spring 1956 he intended to transfer this instrument to someone else and for me electron microscopy (EM) was the logical extension of electron diffraction. So, I could take over that instrument.

At that time most of the EM work in metallurgy was done by the shadowed replica technique. However, Dr.W.Siegfried, who was then head of the metallurgy section, worked on a project for the Union Minéaire du Haut-Katanga about the use of cobalt as an aloying element of austenitic steels (face-centred cubic (fcc) structure). He showed me a paper by Castaing (1) where Guinier-Preston (GP) zones were made visible in Al-4%Cu by transmission EM. (GP zones are precipitates of copper in the form of platelets which are one atom thick). He suggested that this method might be a possibility to see "dislocations" directly and gave me the task to study this possibility in austenitic steels. This was the first time I ever heard about dislocations. (Dislocations are line defects in crystals, elements of plastic deformation. Books about dislocation theory are among others (2,3)). By searching the literature I found the paper by R.D.Heidenreich (4) which is the origin of all aspects of transmission EM, of the preparation technique as well as the theory of the contrast. Heidenreich studied aluminium and prepared his specimens by electro-polishing. As Fig. 2 shows, he placed a pointed cathode against the flat specimen as anode. He let the polishing process act until small holes appeared in the specimen. His crucial problem was then to stop the polishing as quickly as possible since, as soon as the holes opened, the current was concentrated at the edges of the holes and therefore these edges tended to become blunted and the specimen remained thick. In order to determine the exact moment to stop he illuminated the specimen from behind in order to determine the very first moment when the holes appeared.

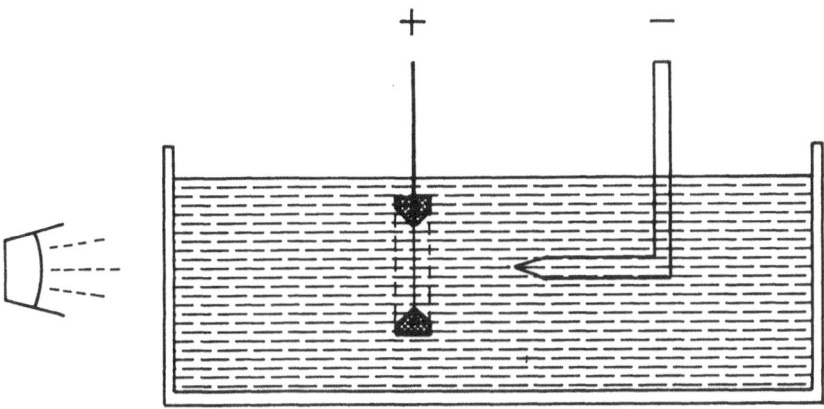

Fig.2:   Heidenreich's electro-polishing arrangement.

For my work, I had first to choose the specimen material and the chemicals for the polishing. For that I asked our electro-chemist Dr.V.Spreter and he gave me an austenitic 18/8 (18%Cr-8%Ni) stainless steel which had been developed specially for the watch industry for its good electro-polishing qualities. The proposed electrolyte was 60% phosphoric acid and 40% sulfuric acid.

In order to understand the difficulties of Heidenreich's preparation technique one has to see the electrical field, its potential surfaces and field lines. As soon as the hole opens, the field lines and with them the current concentrate at the edge of the hole. (My work in the high voltage laboratory

128

had taught me to think in terms of potential distributions). I thought that these difficulties could be avoided in a symmetrical arrangement.

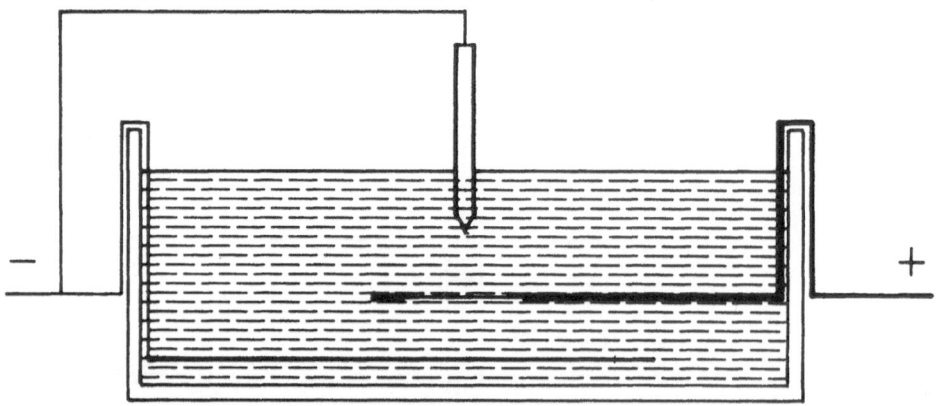

Fig. 3:   Our first trial arrangement.

In the first very crude trial I choose the arrangement of fig. 3. The specimen as cathode was a sheet of stainless steel which was insulated by means of an insulating varnish except for a small window. The anode was a stainless steel sheet on one side and a rod on the other. I gave so little weight to that first trial that I just forgot that it was running. When I remembered it again the whole window was eaten except a small flake at the edge. In the EM this flake proved to be the best specimen for the months to come. (It is to be mentioned that at the beginning of my EM activity I received lots of help from Dr.E.Kellenberger and his collaborators at the University of Geneva. Kellenberger and myself had been studying physics together at the ETH in Zürich in the same semester).

Lots of practical work went afterwards into the refinement of the preparation technique which was done essentially by Mr.P.Fontaine, my laboratory technician at that time. The final arrangement is shown in fig. 4.

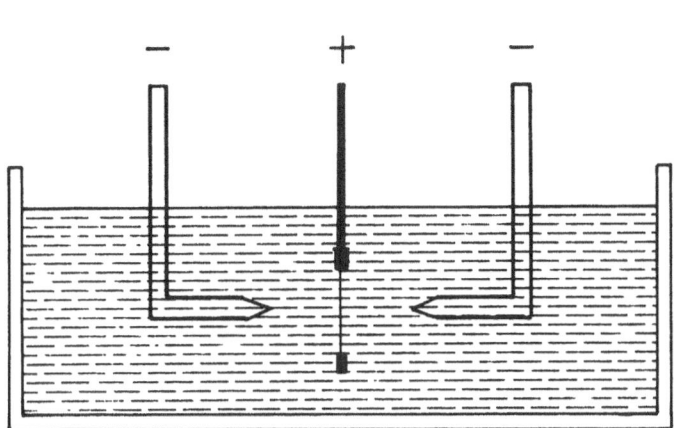

Fig.4:   Our final arrangement.

The spacing of the two pointed electrodes could be changed. The final procedure was then the following: The specimen was in the form of a disk of about 2 cm diameter and 0.2 mm thickness soldered onto a copper wire. The edges of the disk and the wire were insulated by the insulating varnish (Fig. 5a). At the beginning the two pointed anodes were placed at about 2 mm distance from the specimen. The electro-polishing process went on until a hole appeared in the centre (Fig. 5b). Then the spacing of the electrodes was increased and the electro-polishing continued until in the upper part a second hole appeared (Fig. 5c). The process was stopped when the two holes linked up and two tongues were left (Fig. 5d) which could give good specimens. Further problems posed the way of cleaning the specimens and protecting and cutting them for studying them in the EM.

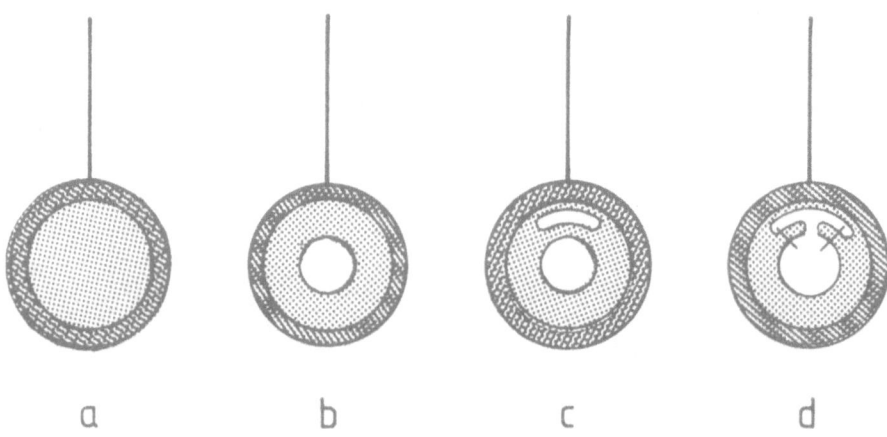

Fig.5: a) Starting situation of the specimen.
 b) Hole in centre after attack. The tips of the electrodes are 2 mm from the specimen.
 c) A second hole opens at larger electrode spacing.
 d) The two holes join. The small tongues are used as specimens.

The EM pictures of the specimens showed features which, for a long time, I did not understand. I saw sections of lines and wondered whether they were on the surface. Every evening I took the pictures home and studied them under the magnifying glass until suddenly on a Friday evening I realized that they must be dislocations inside the metal which became visible because the distortions of the crystal lattice around the line changed the local diffraction conditions and therefore gave the contrast (Fig. 6). Here, my years of preoccupation with electron diffraction paid off for that understanding. (At the beginning it was not obvious that dislocations could be seen by transmission since they represented a displacement by one atomic spacing (about 0.2 nm) while the resolution of the EM 100 was guaranteed to 5 nm).

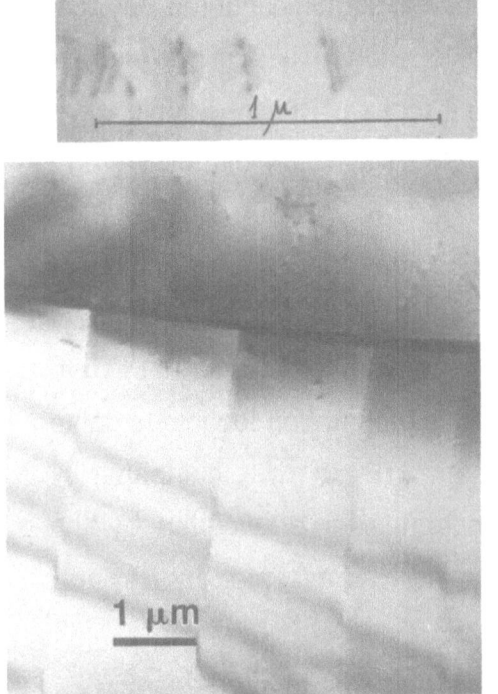
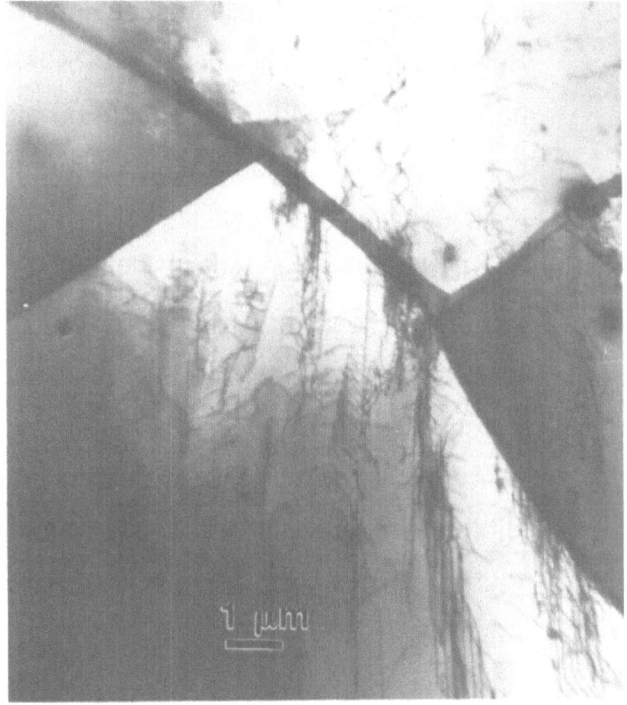

Fig.6: a) Individual dislocation lines in stainless steel lying on a glide
plane and crossing the metal foil (Courtesy: Physical Review).
b) Piled up dislocations on glide planes traversing a grain
boundary (Courtesy: Physical Review).
c) Tangled dislocations in a slightly cold rolled stainless steel
(from the Proceedings of the Stockholm Conference on
Electron Microscopy 1956).

On the following Monday morning I reported the results to our director
Dr.H.Thiemann and proposed to publish them. One of the next days we had
a visit of Dr.C.Crussard, one of the top metallurgists of France. I showed
him my pictures and he asked whether I had seen dislocations. I pointed
them out to him and he was rather astonished and told me that he had
heard two weeks before from Professor Nevil Mott that in Cambridge the
group of Dr.P.B.Hirsch (today Sir Peter Hirsch FRS) had also seen
dislocations by transmission. At first, somewhat disappointed, I considered

132

to drop the plan of publication but then decided to submit a letter to Physical Review anyhow. This discovery of the dislocations was for me really a narrow hit. It could have happened as well that Cr.Crussard could have shown dislocations on my pictures to me. So I was really glad to have at least identified them myself.

In the submitted letter (5) I gave so much importance to the pictures of the dislocations that I just forgot to mention the preparation technique. However the editor wrote back and asked me whether I would not like to describe how the specimens were made. This preparation technique came then to be known as the "Bollmann technique". (By the way, the paper by Hirsch, Horne and Whelan (6) was accepted for publication by the Philosophical Magazine ten days before mine was accepted by Physical Review).

Intermediate Discussion

A very serious mistake of mine in this research was to insist on improving the preparation technique after I had seen that good specimens could already be obtained by the primitive method. One should never loose the aim of the research out of sight and this aim was to study dislocations. A research should, if possible, not be divided into developing of the experimental technique and then, when the technique works, doing research. Each trial should be made so that, if by chance it is succesful, the result is a contribution to the research itself and this result should then be interpreted thoroughly. The aim of the research is often attained before the experimental technique is perfectly working.

An important point in developing a preparation technique is to look not only at the successes but also very carefully at the failures and ask why the trial failed in exactly the way it did. From the "explanation" (which may be wrong) follows a suggestion for a correction. This may lead to a new failure but in another way, the study of which brings again new ideas and finally the technique will work. It is a great error to consider only the successes and to disregard the failures. In this way one might never attain the goal at all.

Another important point, which I realized during this research, was that we are blind for everything which we are not mentally prepared to see. Lots of thinking is necessary for seeing. The dislocations were on my pictures months before I saw them. Heidenreich, in his fundamental paper (4) seven years earlier, mentioned that "... a study of the fine details of the contours in sections plastically deformed under controlled conditions may yield important information concerning dislocations" and, on inspection, after their discovery in Cambridge and Geneva one could see them also on Heidenreich´s published pictures. Today, however, to see dislocations by TEM is an absolute triviality.

In the summer 1956 a meeting of the English electron microscopists was held in Reading (England) where the Hirsch group for the first time showed their ciné-film on moving dislocations in aluminium. I attended the meeting and could show my pictures of stainless steel of different degrees of cold rolling. That was the first time that I met the people of the Cambridge group. From the standpoint of metal physics the comparison between aluminium and stainless steel was very interesting since the former has a specially high "stacking fault energy" while the latter has a very low one. (The stacking fault energy is a material constant which measures the energy difference between the face centred cubic (fcc) and the hexagonal close packed (hcp) structure for the material under the same conditions). The above statement says that aluminium is energetically much farther away from the hexagonal structure than stainless steel. This makes cross slip (change of glide planes) possible in aluminium and the splitting into partial dislocations (formation of stacking faults i.e. one layer hexagonal planes) in stainless steel. It is these properties which make that aluminium stays softer on cold working than stainless steel.

The Cambridge group had prepared their specimens from beaten high purity aluminium (i.e. the extremely thin foils which are used in book binding for imprinting titles on covers). They further dissolved these foils in hydrofluoric acid until small flakes remained as EM specimens. For studying a low stacking fault energy material they had tried a similar technique with equivalent gold foils but without success. So they were rather astonished when I came with stainless steel. As academics they would never have thought to look at such a "dirty" technical material.

In the automn 1956 the International Electron Microscopy Congress was held in Stockholm where Mr.M.J.Whelan (today Dr.M.J.Whelan FRS) showed his film while I presented my stainless steel work. Mr. Whelan asked me there whether I could give them specimens so that they could make a film also on stainless steel. The resulting paper (7) and film seemed to have definitively convinced the metallurgists of the existence of the dislocations because it was done on a technical material and not on a high purity model material.

Recrystallization of Nickel

The next work by TEM was a study of recrystallization of nickel. This work was actually a byproduct of a research on the catalytic activity of nickel by my colleague Dr.H.Schachner. At that time the idea was that the active centres for catalysis were the points where dislocations pierced the metal surface. He asked me to look at the dislocation density of his specimens. (It was found later that the catalytic activity was more due to the nickel oxide on the surface).

Since I already had the material, the preparation technique and the microscope I decided to use the occasion to look at the recrystallization. I rolled a sheet of 0.5 mm thickness for 80% down to 0.1 mm. It was then cut into pieces of the size needed for the preparation of EM specimens (2 x 1 cm). The first was a room temperature specimen, the others were then individually heated in air during 30 min starting with 100 °C in steps of 50 °C up to 400 °C and from their on in larger steps up to 800 °C. The Vickers hardness of each specimen was measured and then the specimen investigated by TEM.

The three observed stages are:

a) Recovery, room temperature to 250 °C (the transition from disordered dislocation tangles to a cellular structure of subgrains),

b) Primary recrystallization, 250 °C to 400 °C (the formation of dislocation free grains separated by high angle grain boundaries) (Fig.7) and

c) Secondary recrystallization, from 400 °C upwards until melting (the increase of the grain size and so the diminishing of the total grain boundary surface and energy).

Fig. 7:   Nickel during primary recrystallization. A practically dislocation
free grain with twin crystals has formed within a polygonized
surrounding. (Courtesy: Journal of the Institute of Metals).

These three stages were well distinct on the hardness curve as well as on the pictures of the internal structure. This work was orally reported on the Fourth International Conference on Electron Microscopy in Berlin 1958 and on invitation at the Symposium on "Application of Thin-Film Techniques to the Electron-Microscope Examination of Metals" held at the Royal Society of London in 1959. This work was the first study on recrystallization by TEM at all (8).

Cobalt Phase Transition

Another work concerned the phase transition of cobalt. Cobalt has the face centred cubic (fcc) structure above about 420 °C and the hexagonal close packed (hcp) one below about 390 °C. The transition is "martensitic" i.e. diffusionless because of the low transition temperature. The fcc structure can be considered as close packing of layers of spheres. Once one layer is given, the position of which we call A, then the next layer has two possibilities which are called B and C. The fcc structure is obtained by a layer sequence of the type ..ABCABC.. while the hcp structure is formed by a sequence ..ABABAB.. . For the transformation from fcc to hcp one can see that every second layer has to be shifted of the type B to C. The transformation is never perfect. The metal is full of "stacking faults" (Fig. 8). Based on the TEM observations I proposed a mechanism which seems to be generally accepted today (9).

138

Fig. 8:   "Framework" of stacking faults in cobalt at room temperature.
This arrangement conserves locally the cubic structure.

Radiation Damage in Graphite

A relatively large work was a study of radiation damage in graphite
which was originally a project for an industry dealing with nuclear reactors.
Since graphite is used as moderator for neutrons in nuclear reactors it is
damaged by neutron irradiation and so stores energy. The stored energy
(Wigner energy) can become so high for high irradiation doses that, on
reheating the material for recovery, the released energy is higher than can

be resorbed by the specific heat so that the temperature of the reactor can run up and damage the reactor core. This happened in a reactor accident in England. The stored energy is represented by crystal defects which can be observed by electron microscopy. The work is described in detail in (10,11).

The available material was reactor graphite which had been irradiated by about 10^{20} neutrons/cm^2. Meanwhile Battelle had acquired a Siemens Elmiskop electron microscope with a guaranteed resolution of 1 nm. The EM specimens were small flakes observable by transmission. Graphite has a layer structure with strong covalent bonds within the layers. The layers are kept together by weak von der Waals forces. In a certain respect graphite can be considered as a stack of two-dimensional crystals. The reactor graphite, which is artificially produced, is highly "turbostratic", i.e. the stacked layers or layer packages are arbitrarily rotated with respect to each other. So, the crystallographic c-axis is in common but the a,b-axes vary in orientation within their plane. This means for the electron diffraction that the superposition of all the reciprocal lattices forms a set of concentric cylinders around the common c*-axis. In contrast to this the superposed reciprocal lattices in a normal polycrystalline material form a set of concentric spheres which lead to the standard Debye-Scherrer ring diffraction patterns when cut by the Ewald sphere. In the case of graphite elliptic diffraction patterns are possible since the cylinders can be cut obliquely by the Ewald sphere (Fig. 9).

The best method for EM observations of graphite proved to be the dark field technique where the pictures were taken in the "light" of diffracted electrons instead of the undeviated ones. By tilting the electron beam so that diffracted beams pass through the axis of the microscope and so are selected by the objective aperture in the focal plane, one picks effectively a few small layer packages out of the whole stack so that the observed specimen is in reality much thinner than that on the specimen grid. In addition the contrast of the dark field picture is much higher than the one of the bright field since in this latter case most of the electrons pass the specimen undeviated and the actual contrast is given only by a small fraction of the electrons. In addition in this case the contrast is determined by all the layers of the specimen and not only by a small selection of them.

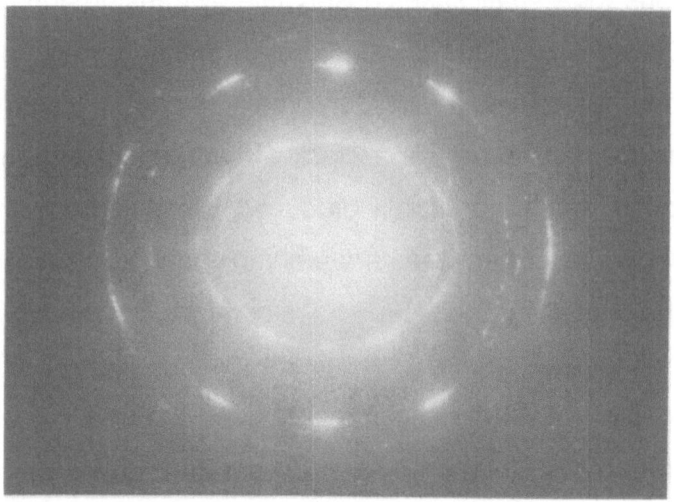

Fig. 9:   Elliptic electron diffraction pattern of graphite.

The first observed specimens showed a pockmark like appearance (Fig. 10) which suggested that the defects were created one at a time as "spikes" (10). Later, however, while working together with Dr.G.R.Hennig at the Argonne National Laboratory (close to Chicago U.S.A.) on a sabbatical leave I was studying graphite single crystals which had been irradiated by 10^{19} neutrons/cm^2 (10 times less than the first specimens). Here it became evident that the carbon atoms had been knocked inidividually out of the graphite crystal. These atoms formed then small interstitial clusters (Figs. 11, 12). The pictures were compared with calculated models based on the kinematical theory of electron diffraction (Fig. 13). The work is reported in (11).

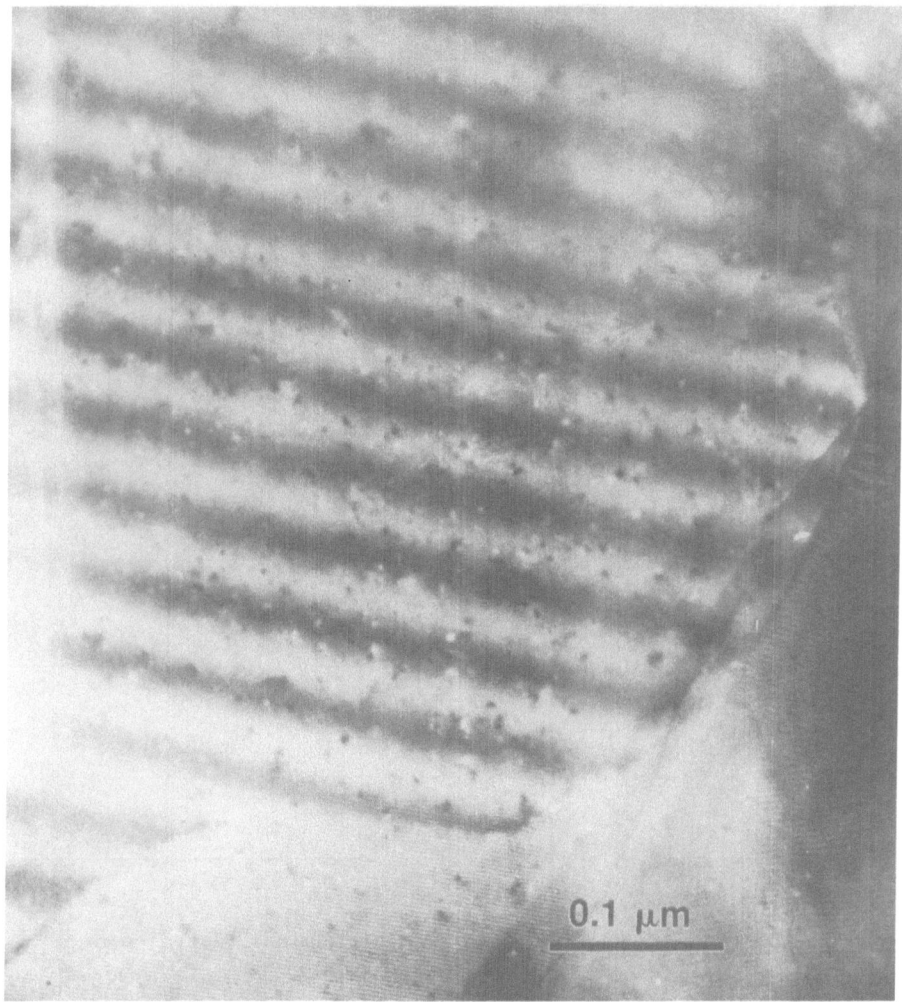

0.1 μm

Fig.10: Dark field picture of reactor graphite (dose: 10^{20} n/cm^2). The moiré stripes are due to the relative rotation of the superposed layer packages. The "pockmarks" are defects due to the irradiation (Courtesy: Philosophical Magazine).

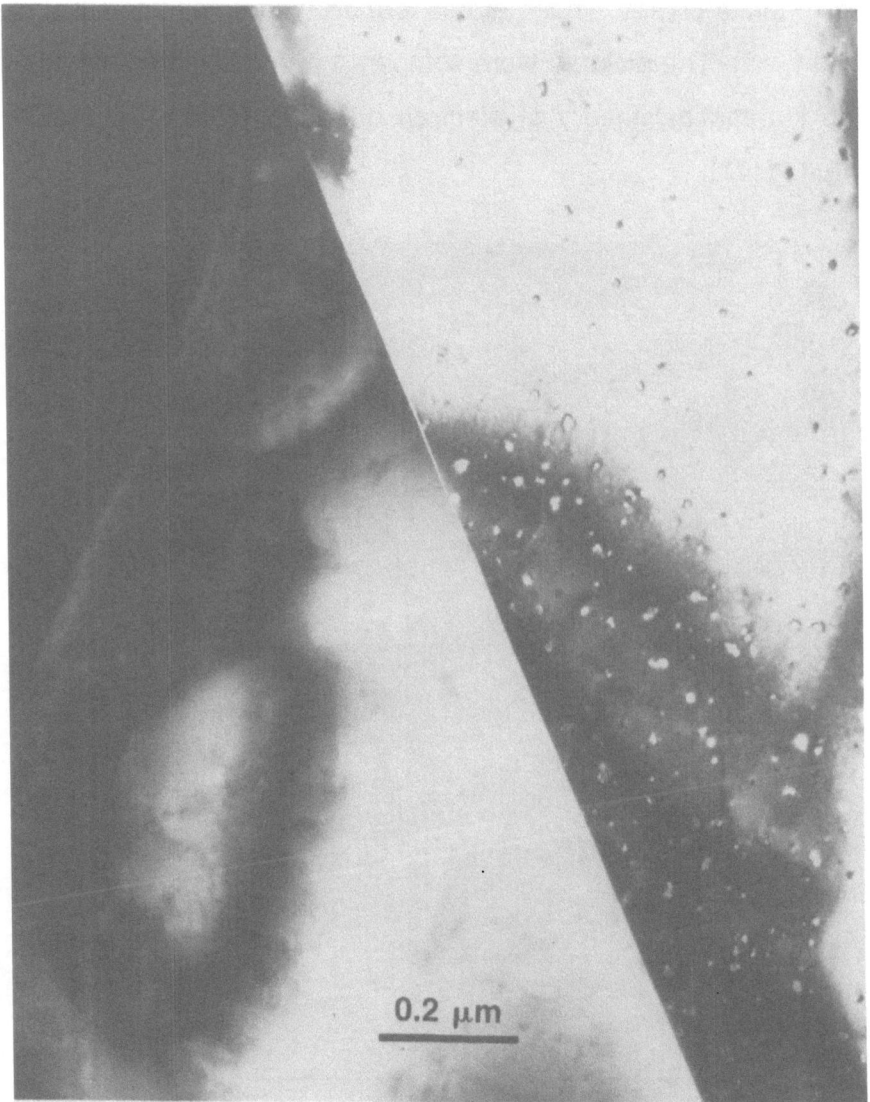

Fig.11: Graphite single crystal (dose 10^{20} n/cm^2, heated to 1580 $^{\circ}$C). This dark field picture is under two different diffraction conditions, (10,0) to the left and (10,1) to the right. The picture is not a composition of two photos but the specimen contained a fold so that by chance the two sides corresponded to the two above indicated diffraction conditions (Courtesy: "Carbon", Pergamon Press).

0.2 μm

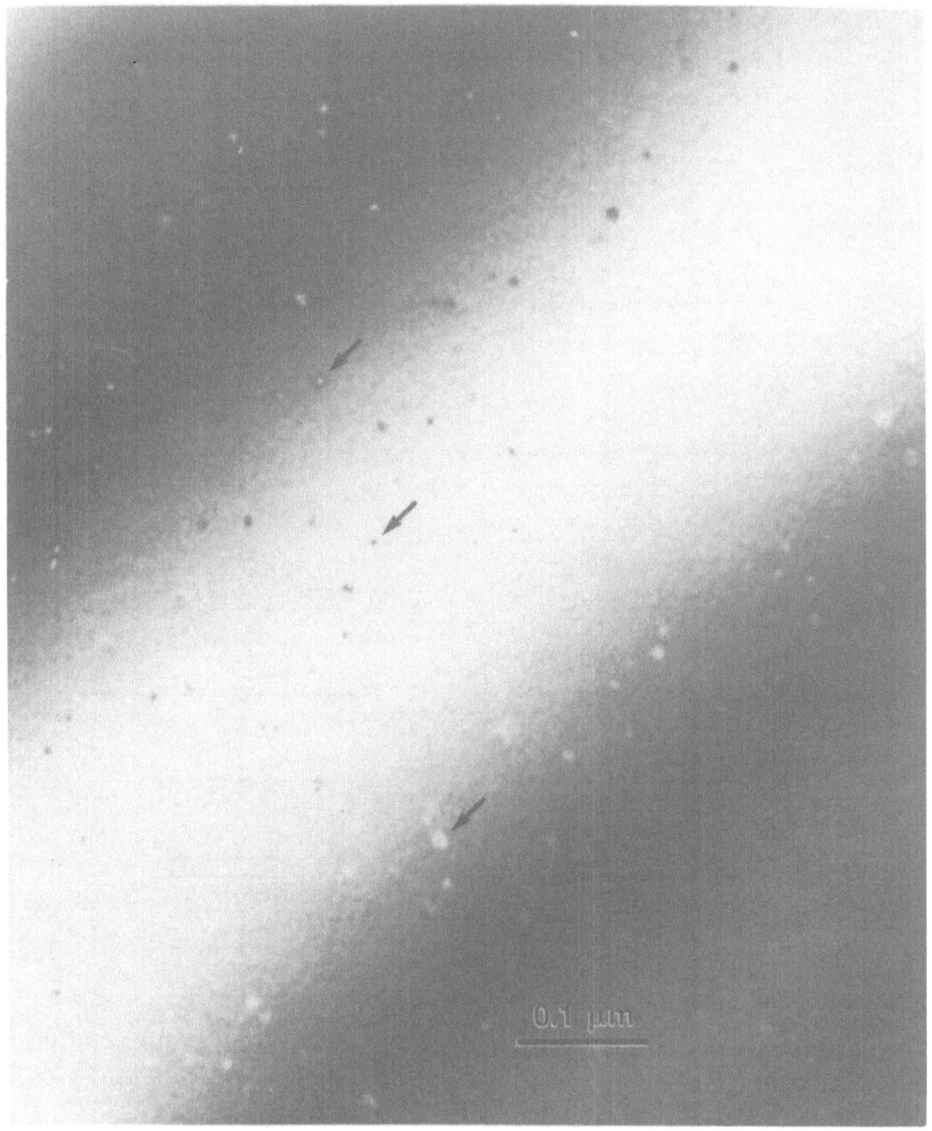

Fig.12: Graphite single crystal ($5 \cdot 10^{19}$ n/cm^2, 1840 ^{o}C), dark field. Extinction band of the (10,1) diffraction. Note the following characteristics: to the left of the band are dark rings with a white centre, in the middle are dark spots and to the right white rings with a dark centre (Courtesy: "Carbon").

144

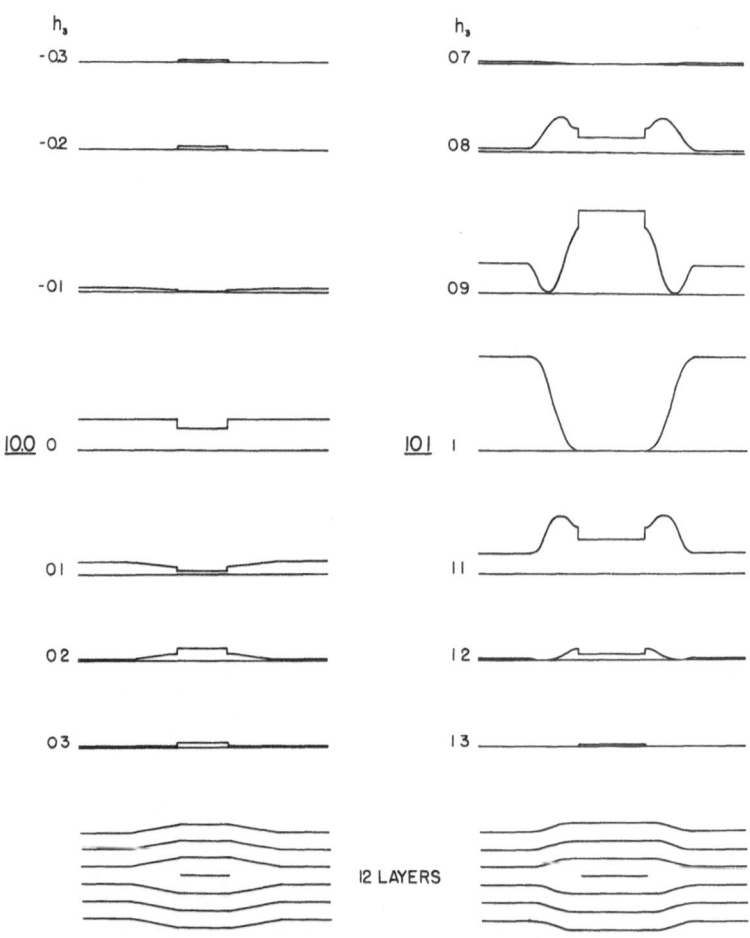

INTENSITY PROFILE OF AN INTERSTITIAL CLUSTER

Fig.13: Calculated intensity profile of an interstitial cluster by means of the kinematical theory of electron diffraction. Note the difference in intensity between the (10,0) and the (10,1) diffraction (see fig. 11) and compare the characteristics of the intensity profile of the (10,1) diffraction between $h_3 = 0.9$ and $h_3 = 1.1$ with fig. 12 (Courtesy: "Carbon").

Theory of Grain and Phase Boundaries

In 1964 I became a member of the Battelle Advanced Studies Center and there my activity shifted then more to the theory of dislocation networks (12) and then the structure of general grain and phase boundaries. This theory has grown out of experience gained by EM observations and was made for understanding the features observable by EM. The essence of this work is given in (13, 14, 15). In this connection one work in collaboration with Dr. H.U.Nissen, at that time in the institute of Professor F.Laves at the ETH, concerned interlamellar phase boundaries in an exsolved alkali feldspar (moonstone) (16) (Fig. 14).

Closing Remarks

This paper is in a sense a brief summary of a whole scientific career. Only few of my 66 publications are mentioned. I am very glad to have been able to give some contribution to the wide field of the study of microstructure in crystalline material. Unfortunately I am unable to mention all the people who have helped me during this whole period and to whom I am really grateful.

Dislocations, stacking faults, precipitates, grain boundaries, etc. are called "crystal defects". With the same attitude one might call fish and the whole aquatic faune and flore "defects of the water" but usually water is considered as the medium in which this faune can live. Correspondingly a crystal can be understood as the medium in which the "faune" of the "defects" can exist and act on the properties and the behaviour of the material. It is just one of the many great merits of the electron microscope to have opened the view on this whole world of the microstructure.

Fig. 14: Feldspar "moonstone" with exsolution lamellae. The diffraction of light on these lamellae gives the blue shine of the gemstone.

References

1. R.Castaing, Rev.mét. **52**, 669 (1955).
2. W.T.Read,Jr.,"Dislocations in Crystals", McGraw-Hill, New Work (1953).
3. J.P.Hirth and J.Lothe, "Theory of Dislocations", McGraw-Hill, New York (1968).
4. R.D.Heidenreich, J.Appl.Phys. **20**, 993 (1949).
5. W.Bollmann, Phys.Rev. **103**, 1588 (1956).
6. P.B.Hirsch, R.W.Horne and M.J.Whelan, Phil.Mag. **1**, 677 (1956).
7. M.J.Whelan, P.B.Hirsch, R.W.Horne and W.Bollmann, Proc.Roy.Soc. **A240**, 524 (1957).
8. W.Bollmann, J.Inst.of Metals **87,** 439 (1959).
9. W.Bollmann, Acta Met. **9**, 972 (1961).
10. W.Bollmann, J.Appl.Phys. **32**, 869 (1961).
11. W.Bollmann and G.R.Hennig, Carbon **1**, 525 (1964).
12. W.Bollmann, Phil.Mag. **7**, 1513 (1962).
13. W.Bollmann, Phil.Mag.**14**, 363, 383 (1967).
14. W.Bollmann, "Crystal Defects and Crystalline Interfaces", Springer, Berlin (1972).
15. W.Bollmann, "Crystal Lattices, Interfaces, Matrices", published by the author, Geneva (1982).
16. W.Bollmann and H.U.Nissen, Acta Cryst. **A24**, 546 (1968).

IV. BIOLOGY AND MEDICINE

History of Electron Microscopy
in Switzerland
Edited by John R. Günter
© 1990 Birkhäuser Verlag Basel

THE CONTRIBUTION OF SWITZERLAND TO THE DEVELOPMENT OF EMBEDDING METHODS IN CYTOLOGY

Eduard Kellenberger
Biozentrum, University
Basel

In the early times of electron microscopy in Geneva (see corresponding chapter) we had a lot of collaborations with physicians and especially with the pathologist E.Rutishauser. David Danon (Israel) carried out his M.D. thesis in this context. In the laboratory for electron microscopy (director E.Kellenberger) at the Institute for Physics (director J.Weigle), Danon developed together with Kellenberger and Bron an ultramicrotome (1) (see chapter by W.Villiger) and new wax mixtures. The first steps represented a progress compared to the former situation (International Conference for Electron Microscopy, Paris, 1950 (1a)), but were quickly superseded by the embedding in methacrylates (2). Of course this method was also immediately introduced in Geneva. It appeared then that bacteria - investigated mainly by E.Kellenberger - were damaged by this process, they apparently "exploded" (5). At the occasion of a conference, this artefact was discussed in detail in a common meeting by A.Glauert (GB), A.Birch-Andersen (DK) and E.Kellenberger (CH) around 1954 and the three decided to develop new embedding resins, as they were convinced that the massive damage to the cells occurred during polymerization. Shortly later this artefact was published by Borysko in USA (3). In this same year, the fruits of the above mentioned discussion appeared as new resins which did not show the explosion artefact. A newly developed polyester resin was introduced by Kellenberger (CH), Schwab (CH) and Ryter (CH) (4), an epoxy resin by Birch-Andersen (DK) (5a) and another (as a variant of Araldite, R Ciba Geigy) by Glauert (GB), Rogers (GB) and Glauert (GB) (6).

All these new resins showed no such polymerization artefact and were therefore used world-wide. The methacrylates were rapidly replaced by them, except for a few special applications. Because the polymer chemist W.Schwab, who cooperated with the Geneva group, could produce his polyester resins ("Vinox") only on the laboratory scale, various lots of production were not always identical. Therfore an industrial supplier was sought and found in Germany ("Vestopal" (7)). Other problems appeared later with Vestopal because the distributing company did not find a laboratory to perform the required tests for adapting the concentrations of initiator and catalyser at the different production lots. These circumstances led to the present use of epoxy resins in practically all electron microscopy laboratories.

For embedding with these resins the tissue has to be dehydrated in ethanol (or other water miscible organic solvents) before being infiltrated with the liquid, yet unpolymerized resin. W.Bernhard, a Swiss citizen from Worb who, after his studies in Berne, was in charge of electron microscopy at the Cancer Institute in Villejuif, France, had the idea that without these solvents an improved conservation of the structure should be possible, if the embedding medium itself was polar i.e. similar to water. With regard to cytochemical investigations therefore monomers of resin that were "watersoluble" or miscible with water were sought (8). Another Swiss, W.Stäubli, worked temporarily with W.Bernhard at that time. Coming from the Ciba AG in Basel, he had the required access for the development of a watersoluble epoxy resin (Durcupan) (9, 10), which was used for some time until it was found that the similarity with water might be less important than previously thought (11, 12).

Later work of the Basel group was stimulated by a cooperation with Sjöstrand (USA) who conducted UV-polymerizations with hydroxy-propyl-methacrylate (13). Kellenberger, Villiger and Carlemalm (14) developed resins based on methacrylate with crosslinking, which allowed handling at low temperatures and conserved polar groups after their polymerization, in contrast to most of the former resins. They presented two types of resins (polar and apolar) on a similar chemical basis and with similar properties (15, 16). These resins are produced by a German firm as "Lowicryls" R. They permit to study the influence of polarity without having to change the type of resin as was previously necessary. These resins proved to be also very suitable for Immuno- and Lektin-labelling (11, 17, 18). The group around Kellenberger furthermore developed a resin yielding a negative contrast because of its tin content (19). Both the Sn-resin and the Lowicryles were developed also with regard to the observation of completely unstained tissue by means of ratio contrast in the STEM (20).

154

References

1. D.Danon and E.Kellenberger, Arch.des Sci. Genève 3, 169 (1950).
1a.Comte rendue du premier congrès de microscopie éléctronique, Paris 1950. Ed. rev.d´optique, 1952.
2. S.B.Newman, E.Borysko and M.Swerdlow, J.of Res.NBS 43,182 (1949).
3. E.Borysko, J.Biophys.Biochem.Cytol. 2, 3 (1956).
4. E.Kellenberger, W.Schwab and A.Ryter, Experientia 12, 421 (1956).
5. A.Birch-Andersen, O.Maaloe and F.S.Sjöstrand, Biophys.Biochem.Acta **12**, 395(1953).
5a. O.Maaloe and A.Birch-Andersen, in: Bacterial Anatomy, p.261. The University Press, Cambridge (1956).
6. A.M.Glauert, G.E.Rogers and R.H.Glauert, Nature 178, 803 (1956).
7. A.Ryter and E.Kellenberger, J.Ultrastruct.Res. 2, 200 (1958).
8. E.H.Leduc, V.Marinozzi and W.Bernhard, J.Roy.Micr.Soc.81,119(1963).
9. W.Stäubli, C.R.Acad.Sci. 250, 1137 (1960).
10. W.Stäubli, J.Cell Biol. 16, 197 (1963).
11. M.Dürrenberger, W.Villiger, B.Arnold-Schulz-Gahmen, B.M.Humbel and H.Schwarz, to appear in Vol.3 of the Hayat series on gold labelling.
12. E.Kellenberger, in: Cryotechniques in biological electron microscopy", R.A.Steinbrecht and K.Zierold, eds., Springer, p.35 (1987).
13. F.S.Sjöstrand and F.Kretzer, J.Ultrastruct.Res. 53, 1 (1975).
14. E.Kellenberger, E.Carlememalm, W.Villiger, J.Roth and R.M.Garavito, Low denaturation embedding for em of thin sections, publ.by: Chem. Werke Lowi, GmbH, Postf. 1660, D-8264 Waldkraiburg.
15. E.Carlemalm, R.M.Garavito and W.Villiger, J.Microsc. 126, 123 (1982).
16. J.-D.Acetarin, E.Carlemalm and W.Villiger, J.Microsc. 143, 81 (1986).
17. J.Roth, The protein A-Gold (pAg) technique, in: Techn. in Immunocyto-chem., I, G.R.Bullock and P.Petrusz eds., Academic Press, 107 (1982).
18. J.Roth, J.Microsc. 143, 125 (1986).
19. J.-D.Acetarin, W.Villiger and E.Carlemalm, J.EM Techn. 4, 257 (1986).
20. E.Kellenberger, E.Carlemalm, W.Villiger, M.Wurtz, C.Mory and Ch. Colliex, Ann.New York Acad. Sci. 483, 202 (1986).

History of Electron Microscopy
in Switzerland
Edited by John R. Günter
© 1990 Birkhäuser Verlag Basel

APPLICATIONS OF TRANSMISSION ELECTRON MICROSCOPY IN BIOLOGICAL SCIENCES: PREPARATION PROCEDURES, CONTRAST MODIFICATIONS AND ULTRASTRUCTURAL CYTOCHEMISTRY

Alain Gautier
Center of Electron Microscopy
University of Lausanne

"La Recherche comme
aventure est joie...."
(W.Bernhard, in: Biol.
Cell. **33**, xxx (1978))

I. Introduction

It would be exaggerated to state that the introduction of three transmission electron microscopes in three Swiss universities (Universities of Berne and Geneva, Federal Institute of Technology in Zürich) in the years 1946-48 had invoked enthusiasm in the university circles specialized in biology and medical sciences. Fortunately enough, however, three scientists realized immediately the impact of this new method of investigation: Professor Frey-Wyssling in Zürich (see chapter by K.Mühlethaler), Professor Weiglé in Geneva (see chapter by E.Kellenberger) and a little later Professor Lehmann in Berne (see chapters by J.R.Günter and by E.Kellenberger on the "Bernese Connection"). Furthermore, the importance for our country of the pioneering work of Dr.W.Bernhard pursued in the Cancer Research Institute in Villejuif, France, must be recalled (see chapter by E.Kellenberger on the "Bernese Connection"): he has indeed contributed considerably to the formation of many Swiss scientists, whose work will be discussed below.

Albert Frey-Wyssling (1900-1988)

Professor Frey-Wyssling, director of the Institute for General Botany at the Federal Institute of Technology in Zürich, was already well-known before the war for his concepts of the fine structure and physiology of plants, especially of plant cell physiology. He had enjoied a remarkable formation in mathematics, in chemistry with Staudinger, and in physics and crystallography. Prepared such, he could have become an eminent physico-chemist, but he was attracted already at an early age by the cell physiology of plants. Using crystallographic techniques, such as polarized light microscopy and X-ray analysis, he rapidly founded the concepts of what was called "micellar theory of protoplasms", a theory which he applied in particular to the study of cellulose walls, of starch grains and of chloroplasts (6). "Il avait compris très tôt que, pour avancer dans cette voie très peu prospectée jusqu'alors, il fallait explorer la *terra incognita* qui se cachait au-delà (ou en-dessous) du pouvoir séparateur du microscope photonique et étudier les infrastructures dont l'existence lui paraissait la plus vraisemblable" ("He understood at an early date that for an advancement on this little known way, the unknown world hidden below the resolving power of the light microscope had to be explored and the infrastructures studied, the existence of which seemed to him most probable") (L.Fauconnet, private communication).

Starting from these ideas, he intuitively realized quickly that the new conquests of experimental physics in the domain of electron optics might furnish the instruments he was missing. Therefore, from 1944, he got into contact with G.Induni, engineer at Trüb, Täuber & Cie AG, to plan with him the realization of a first Swiss TEM. In 1944, he sent one of his young coworkers, K.Mühlethaler (see his chapter), to Trüb, Täuber in order to realize the first electron micrographs of biological material taken in our country.

Jean Weiglé (1901-1968)

In 1945, Professor Weiglé was director of the Institute of Experimental Physics at the University of Geneva; however, his scientific interests were already oriented towards biophysics and molecular biology, sciences to the development of which he was later going to contribute considerably in the United States, cooperating with Dellbrück (see chapter by E.Kellenberger). This is why he immediately put a young assistant, E.Kellenberger, in charge of the new electron microscope installed in his laboratory - not without problems, since the Institute of Physics was still in the old and fragile central buildings of the University - and trained him in the new disciplines to which he was going to devote himself.

158

Fritz E. Lehmann (1902-1970)

Professor Lehmann was, at the of the war, well-known for his numerous studies on embryology and on structures of lower eucaryotes. He realized as one of the first the possibilities which the electron microscope might offer for his work. For this purpose he used the prototype Trüb, Täuber instrument installed in the laboratory of Professor W.Feitknecht in the Institute for Inorganic Chemistry at the University of Berne.

Wilhelm Bernhard (1920-1978)

As a young physician, just after having received his diploma, Wilhelm Bernhard from Worb (Berne) left for Paris in 1947 for studies in pathology. He soon became the most important collaborator of Professor Charles Oberling, the famous French cancerologist who had just been nominated as head of the Institute for the Scientific Research of Cancer 'Gustave Roussy' in Villejuif. The latter intrusted him with the directorship of his new laboratory for electron microscopy equipped with a Trüb, Täuber instrument, as were the Swiss institutes. Bernhard soon built up a research unit devoted to experimental cancerology, but also to studies of the ultrastructure of normal cells - in order to have results for comparison with pathological cells - as well as to virology. Oberling had indeed been one of the first to support the hypothesis of the possible role played by viruses in the origin of cancer. The disposition of Bernhard led him not only to concentrate on biological material as such, but also on the quality of its preparation. It is this complimentarity of his studies, on one hand on 'biological' aspects, but on the other hand on the aspects 'conservation of structure', which essentially characterizes Bernhard's leading influence on ultrastructural biology in France, his adopted country, as well as on the parallel development of research effected in Switzerland by many scientists who had past longer of shorter periods of study with Bernhard and retained friendly relations and intense cooperations with him.

The influence of these four scientists on the later development of ultrastructural research applied to biology in our country need not be discussed. Their scientific rigour and interdisciplinary plurality and finally their imagination had a decisive influence on the formation of young scientists who took their succession in the various laboratories of our country, such as Mühlethaler, Kellenberger, Gautier, Rouiller, Stäubli, Koller, etc.

II. The Antiquity (1946-1961)

Geneva

With his scientific rigour and dynamics inherited from his formation as physicist, Eduard Kellenberger succeeded in the progress of gathering around the electron microscope installed in the Physical Institute of Geneva a group of young scientists from biological sciences interested in this new method of investigation. To name a few among them: Werner Arber, Antoinette Ryter, Janine Séchaud, Charles Rouiller, Gérard de Haller, David and Mathilde Danon. No aspect whatsoever of specimen preparation techniques escaped their critical views and their imagination: first the study of isolated elements, then replicas, and finally ultramicrotomy (see article by E.Kellenberger). Let us recall e.g. the wide distribution of the 'R+K' fixation (Ryter and Kellenberger (20)), the applications of which were so numerous, the development of new embedding materials based on polyesters, and further that the problems of conserving the fine structure of nucleic acids quickly became their main preoccupation.

In 1958, Charles Rouiller was nominated as director of the Institute for Histology of the University of Geneva. After his medical studies in this town, he had been assistant of the Professor of Pathology E.Rutishauser; already in this period he had tried to apply the replication technique to the study of bone in the laboratory of Kellenberger. Then, from 1954 to 1958, he had stayed with Bernhard in Villejuif, later at the Collège de France under the direct supervision of Oberling. During his cooperation with Bernhard he had discovered the existence of an unknown kind of components of cells, the 'microbodies', known today under the name of 'peroxysomes' (19). Since his arrival in Geneva, he surrounded himself by an efficient group working essentially on problems of histology and pathology (see chapter by Y.Kapanci). His intense continuous interests in problems of specimen preparation and especially in the systematization of records obtained from ultrastructural studies should be stressed as well. He also knew to give the medical research, first in Geneva and then in Switzerland, the necessary base for becoming a useful and renowned scientific subject.

Zürich

At the Federal Institute of Technology (ETH), K. Mühlethaler developed rapidly an important activity of research in the domain of vegetal cells and in particular of their walls with the first instrument he had obtained (see also articles by Mühelthaler and Moor). It should be stressed that the research undertaken by the group he had formed rapidly developed towards problems of the use of low and ultra-low temperatures in biology. Here, a large field of techniques evolved by the efforts of Mühlethaler's group, including in particular H.-R.Müller and H.Moor (see his chapter).

Also at the University of Zürich, various research groups from the faculty of medicine took advantage of the possibilities offered to them by the laboratory at the ETH. Under the direction of Professor A.von Albertini and with the experience gained by H.U.Zollinger through his cooperation with W.Bernhard, A. Vogel as coordinator put the various groups rapidly into operation, the majority of their problems being centered around bio-medical applications (15). Others however also concerned specimen preparation techniques. It should be noted in particular that Vogel worked since 1957 on the embedding techniques with metacrylates at low temperatures, thus paralleling research by the group of Mühlethaler.

<u>Berne</u>

Since the fifties, Professor Lehmann in cooperation with his assistant R.Weber and an Italian colleague from the University of Bari, Professor A.Bairati, investigated several techniques dedicated to the studies of the most primitive organisms, such as *Tubifex* (12) and *Amoeba* (1).

These first laboratories were soon joined by those of the Universities of Lausanne and Basel.

Lausanne

A new laboratory for electron microscopy was created in 1955 at the University; I have had the chance to become its head, after having studied for one year with E.Kellenberger and for five years with W.Bernhard in Villejuif. The small group I headed was occupied with problems of specimen preparation of blood cells (under the direction of the remarkable hematologist Dr.Robert Feissly (5)), the study of the influence of fixation on the structure of the lung (3), then the demonstration of intracellular glycogen by modification of contrast (section-staining) (11).

Basel

At the Institute of Anatomy of the University, Professor G.Wolf-Heidegger created a laboratory for electron microscopy since 1956, also equipped with a Trüb, Täuber instrument, scientifically headed by M.Thürkauf and with W.Villiger as technical director. This group was particularly interested in freeze-drying techniques as well as in a great number of applications.

An important meeting of the International Society for Cell Biology was held in Berne in 1961, devoted essentially to the study of cellular ultrastructure; the communications of this symposium (23) have been the subject of a publication in 1962. This date seems to me to be particularly important for Switzerland. As a matter of fact, this Symposium has shown to a wide scientific audience that transmission electron microscopy in biology had obtained its standing in own rights, that it was really feasible and had thus become a routine technique.

III. The Middle Ages (1962-1979)

Electron Microscopy in biology, a routine technique ? Yes, in a certain sense. It is true that this period was marked by the blooming of descriptive articles strictly morphological, in our country as well as all over the world. However, a number of techniqes still had to be improved, and it is to this task that several groups of Swiss scientists devoted themselves.

Geneva

At the Institute of Physics, Kellenberger continued his activities until 1970 (see his chapter) with various coworkers. With J.Dubochet, e.g., he continued to develop dark field observation methods of isolated biomacromolecules.

Also at the Institute of Histology, later at the Department of Morphology, Rouiller continued his detailed observations, in particular in the field of normal and pathological hepatology; studies that have been summarized in a joint publication for which he takes the responsibility (18). Unfortunately, Charles Rouiller died in 1973, after he had been elected as chancellor of the University. His research was continued and further developed by his former assistant and later successor, Professor L.Orci (see his chapter). He further developed in particular, together with A.Perrelet and following the work of the Zürich group of H.Moor, the techniques of freeze-etching in histological research (17). He grouped around him a large number of valuable coworkers, such as Perrelet, Forssmann, Matter, etc., with whom he participated in the development of numerous other methods.

Zürich

At the Institute for Botany of the Federal Institute of Technology, the group directed by K.Mühlethaler also continued its research, in particular in the field of low temperatures. This fundamental work led H.Moor and coworkers to the invention of a new preparation method, the cryo-fractioning, which attracted world-wide interest and dissemination (see chapter by Moor). The instruments developed by Moor and his group were later on commercialized by the Balzers AG in Liechtenstein.

At the University, a new laboratory was created in 1963 by E.Weibel. He had started to develop methods of light optical morphometry at the Columbia University, USA, in 1959, then of their applications to transmission electron microscopy in 1961 at the Rockefeller Institution; in Zürich he demonstrated together with G.Kistler that stereology was a field with a prosperous future, the applications of which became innumerable (see chapter by E.Weibel).

Basel

New laboratories were created outside the University, particularly in the big chemical industries.

In the Ciba laboratories, W.Stäubli continued work on embedding media based on water soluble araldites, a field he had started to work in at

Villejuif with W.Bernhard. His results proved rapidly to be very useful and the water soluble araldite Durcupan became to be used in many laboratories.

At Hoffmann-La Roche, J.-P.Tranzer was engaged in 1963 as head of the laboratory of ultrastructural research. This young French scientist (1928-1974) had studied chemistry in Mulhouse (France), worked in industry, stayed for further studies in the United States and finally finished his medical studies in Strasbourg (France). Further studies aimed at specialization followed in Villejuif with Bernhard. Since 1966, he has published in Basel a large number of papers on ultrastructural cytochemistry - we only cite two of them (24, 25). One of his prime interests was the localization and preservation of biogenic amines. He knew to teach his Swiss colleagues the rigour required for applying cytochemistry. Unfortunately, he died after a very short period of severe illness in 1974.

In automn 1971 the University of Basel inaugurated its completely new Biocenter. Kellenberger, who had been called from Geneva in the previous year, directed there in collaboration with W.Arber the Microbiology Division. He grouped around him a number of valuable scientists, such as Carlemalm, Dubochet, Villiger and Wurtz and they developed new techniques for specimen preparation and for the observation of isolated particles, but also new embedding materials. The Lowicryl resins, essentially developed by Carlemalm, became materials of current usage, with particularly favorable properties for their application in immunocytochemistry (see chapter by Kellenberger on Embedding Methods). These properties were clearly demonstrated by J.Roth, when he worked in Geneva in the laboratories of L.Orci, and were further refined when he, too, was called to the Biocenter. The installation of a STEM at the

Biocenter should also be pointed out (see chapter by A.Engel), with all the consequences this group of scientists drew from it, both in the domain of high-resolution and that of quantitative measurements in molecular biology.

Lausanne

The group at the Center of Electron Microscopy (Centre de Microscopie Electronique - CME) became more and more occupied with refining cytochemical techniques during these years, with the help of several scientists, among whom were many Italians (V.Marinozzi (14), L.Lombardi (13), F.Minio, L.Okolicsanyi, G.Prenna, F.Bonvicini (2), etc.). Most of these techniques were applied to various problems of hepatology (see chapter by Y.Kapanci). Since 1966, following a suggestion of W.Bernhard, the main interest was concentrated onto methods for studying nucleic acids. Through collaborations with R.Cogliati (4), M.Schreyer, J.Fakan and others we could demonstrate successively various reactions, studies which were inspired by former work of Robert Feulgen on the localization of DNA in light microscopy. The first results on the 'Feulgen-type' reaction with ammine complexes of osmium were published in 1973 (3), as was a review of all the techniques of ultrastructural cytochemistry concerning the *in situ* investigation of DNA in 1976 (7).

Berne

E.Weibel became director of the Institute of Anatomy of the University in 1966. He transferred his laboratory from Zürich to Berne and developed considerably his research of theoretical and practical ultrastructural morphometry (see his chapter), the importance of which has not ceased to increase on an international level.

Fribourg

A small group around Professor P.Sprumont developed at the University various techniques for ultrastructural cytochemistry, by studying in particular the use of metal ions (21, 22).

Vevey

Finally we should also report the methodological studies performed in the laboratories of Nestlé, first in Vevey and later near Lausanne. Already in 1959, H.-R.Müller, a former coworker of Frey-Wyssling in Zürich who had considerably contributed to the development of cryopreparation techniques, created a laboratory for the prepartion of specimens for electron microscopy, the observations being performed at the CME in Lausanne. His research was essentially directed at the preparation of primary products for the food industry (16) for ultrastructural analysis. His successors as head of the laboratory, which had in the meantime become equipped with a TEM in 1966, were first H.Bauer, also a former pupil of Frey-Wyssling, then Marc Horisberger, who is now director of the Nestlé Research Center in Lausanne. Of particular importance is the invention by Horisberger and his coworkers (8) of the colloidal gold technique, its development in ultrastructural cytochemistry (lectins) and in immunocytochemistry (9, 10). Fifteen years later, it has become evident that the colloidal gold technique has become an indispensable tool in all immunocytochemical research.

IV. Contemporary Times (1980-1990)

This last decade has seen an actual explosion of instrumental possibilities in the domains of electron optics and related fields. Due to this, we can not yet present a review appreciating the major methodological contributions to biological applications.

High-resolution STEM, TEM or SEM equipped with accessories for elemental analysis by X-rays (XRMA), TEM with analytical accessories for elemental analysis by electron energy loss (EELS), or apparatus for cryo-observation, tunnelling microscopes, but also computerized instruments for analysis as well as for image processing, for specimen preparation at very low temperatures, at high speed or/and high pressure ... - such a list can certainly not be exhaustive. The number of possibilities for ultrastructural analyses actually available to the scientist in experimental biology or in medical sciences exceeds the imagination.

However, it may also not be overlooked that acquiring such instruments as well as the need of having a group of pluridisciplinary scientists in an inspiring atmosphere for cooperation around each of them, require considerable financial means. It is therefore not surprising that the principal studies done in our country are concentrated in a relatively small number of laboratories.

It is fortunate to know that the scientists in these centers pursue their work today with the same enthusiasm and as efficiently as it has been the case since 1946. We cite here only a few of them - this list is certainly not exhaustive as well - the publications of whom have attired the interest of specialists:

In <u>Basel</u>: U.Aebi, Ch.Brack, M.Dürrenberg, A.Engel, E.Kellenberger, J.Meyer, R.Reichelt, J.Roth, W.Villiger.

In <u>Berne</u>: L.Cruz-Orive, P.Gehr, E.Hunziker, E.R.Weibel.

In <u>Geneva</u>: L.Orci, A.Perrelet, J.D.Vassalli.

In <u>Lausanne</u>: J.Dubochet, S.Fakan.

In <u>Zürich</u>: H.Gross, G.Kistler, T.Koller, M.Müller, H.Moor, A.Stasiak.

It is obvious that the possible applications of ultrastructural research to life sciences at present concern all disciplines, from molecular biology to anatomy, from ecology to immunology, from genetics to AIDS research.

Acknowledgements

I wish to thank all my colleagues who have supplied valuable information, in particular L.Fauconnet, H.Kuhn, G.Richards, P.Sprumont, W.Villiger, A.Vogel, R.Weber, E.R.Weibel and O.Zwillenberg.

References

1. A.Bairati and F.E.Lehmann, Pubbl.Staz.Zool.Napoli **23** Suppl., 193 (1951).
2. F.Bonvicini, A.Gautier, D.Gardiol and G.-A.Borel, Lab.Invest. **38**, 487 (1978).
3. M.Campiche, J.Ultrastruct.Res. **3**, 302 (1960).
4. R.Cogliati and A.Gautier, C.R.Acad.Sci.(Paris) **276D**, 3041 (1973).
5. R.Feissly, A.Gautier and I.Marcovici, Rev.Hémat. **12**, 397 (1957).
6. A.Frey-Wyssling, "Submicroscopic Morphology of Protoplasm" (2nd ed.), Elsevier Publ.Comp., Amsterdam (1953).
7. A.Gautier, Intern.Rev.Cytol. **44**, 113 (1976).
8. M.Horisberger, J.Rosset and H.Bauer, Experientia **31**, 1147 (1975).
9. M.Horisberger and J.Rosset, J.Histochem.Cytochem. **25**, 295 (1977).
10. M.Horisberger and M.Vonlanthen, J.Microsc. (Oxford) **115**, 97 (1979).
11. G.Jean and A.Gautier, C.R.Acad.Sci.(Paris) **253**, 2274 (1961).
12. F.E.Lehmann, Rev.Suisse Zool. **57** Suppl., 141 (1950).
13. L.Lombardi, G.Prenna, L.Okolicsanyi and A.Gautier, J.Histochem. Cytochem. **19**, 161 (1971).
14. V.Marinozzi and A.Gautier, J.Ultrastruct.Res. **7**, 341 (1962).
15. K.Mühlethaler, A.F.Müller and H.U.Zollinger, Experientia **6**, 16 (1950).
16. H.-R.Müller, Milchwiss. **19**, 345 (1964).
17. L.Orci and A.Perrelet, "Freeze-Etch Histology", Springer, Berlin (1975).
18. Ch.Rouiller, in: "The Liver" (Ch.Rouiller, ed.), Academic Press, New York and London, Vol. **I** (1963) and Vol. **II** (1964).
19. Ch.Rouiller and W.Bernhard, J.Biophys.Biochem.Cytol. **2** Suppl., 355 (1956).
20. A.Ryter and E.Kellenberger, Z.Naturforsch. **13b**, 599 (1958).
21. P.Sprumont and J.-P.Musy, Histochemie **26**, 228 (1971).
22. P.Sprumont, J.-P.Musy and V.de Blasi, Acta Histochem.**42**, 285 (1972).
23. "The Interpretation of Ultrastructure" (R.J.C.Harris, ed.), Academic Press, New York and London (1962).
24. J.-P.Tranzer, M.Da Prada and A.Pletscher, Nature **212**, 1574 (1966).
25. J.-P.Tranzer and J.G.Richards,J.Histochem.Cytochem.**24**,1178 (1976).

History of Electron Microscopy
in Switzerland
Edited by John R. Günter
© 1990 Birkhäuser Verlag Basel

THE CONTRIBUTION OF ELECTRON MICROSCOPY TO NORMAL MORPHOLOGY AND PATHOLOGY RESEARCH IN SWITZERLAND

Yusuf Kapanci
Department of Pathology, Faculty of Medicine
University of Geneva

Introduction

The aim of this article is to recall some major contributions of Swiss electron microscopy (EM) investigators in the fields of histology and pathology. This review does not claim to cover all investigations done in Switzerland, but to cite a few outstanding ones since the early fifties. We are conscious that possibly some colleagues are forgotten, or citation of some of their important contributions is missing in the following chapter. We hope that they will forgive us.

A historical overview on the utilisation of EM in different Swiss institutes of biology is reported in other papers published in this volume. We cannot however omit to recall Jean Weigle from Geneva who, even before 1950, put the EM at the disposal of biological research. Nevertheless his main contribution in this field was after his move to the USA; his outstanding studies on bacteriophages constituted then a pioneer work which led eventually to the outburst of the molecular biology era. During this period he collaborated very closely with E.Kellenberger from Geneva (see his chapters). This latter investigator has certainly been one of the most prominent figures among Swiss and international EM investigators (1). His contributions are summarized elsewhere in this book. However we cannot overlook the fact that he trained investigators such as Charles Rouiller and Alain Gautier; he collaborated with outstanding pathologists such as Erwin Rutishauser and co-workers, contributing thus to the advancement of ultrastructural research in morphology.

Two other names to recall are W.Bernhard from Berne and A.Vogel from Zürich. The first pursued his career in Paris and published his most important studies on viral carcinogenesis in collaboration with Ch.Oberling (2). Nevertheless he kept a continuous relationship with Swiss scientists, particularly with Ch.Rouiller from Geneva, and thus contributed to the development of EM research in this country. A.Vogel's earliest important studies are technical: he described a method for rapid embedding in butylmethacrylate and polymerization by X-rays (3). Later all through his career he participated in numerous studies in pathology: in collaboration with H.U.Zollinger on kidney pathology, with A.von Albertini and J.R.Rüttner on liver diseases and on fundamental cancer research, and with U.W.Schnyder (4) on skin diseases.

In this manuscript we shall discuss first the main contribution of EM to normal morphology, then to pathological conditions. Evidently, there will be some overlaps between the two fields. Hence we shall divide the following paragraphs into research done in histology (or morphology) departments and in pathology departments.

A. Histology - Morphology

Among Swiss EM investigators, Charles Rouiller, former head of the *Department of Morphology in Geneva*, is probably one of the best known histologists. His very first contribution has been in the field of ultrastructural morphology of bone while he was working with E.Rutishauser in the Institute of Pathology of Geneva (5). However we owe Rouiller's greatest scientific contribution and his international figure to his studies on liver. In 1963, he edited an Academic Press monograph (6), in which one can find a most relevant ultrastructural study of the liver, and particularly a functional approach through morphology of the hepatocyte. In later studies, Rouiller emphasized the differences between endothelial and Kupffer cells (7-10).

During the late period of his activities, Rouiller's studies were oriented towards adipous tissue, kidneys, and more particularly endocrine pancreas (11). Together with his valuable pupil and successor as the head of the Department of Morphology of the Geneva Medical School, Lelio Orci, he published some very fundamental studies on islet cells of the pancreas and on diabetes (see Orci's publications). During his activity as professor of histology, Ch. Rouiller collaborated with world-wide known personalities such as M.J.Karnovsky, G.Majno (see his publications), and furthermore trained famous research fellows such as W.G.Forssman, A.Matter and Lelio Orci.

Lelio Orci, who after the sudden death of Rouiller in 1972 became the head of the department of morphology in the Geneva School of Medicine, is an outstanding EM investigator. His contributions concerning insulin biosynthesis and secretion, the distribution of islet cells in the pancreas and related endocrine cells in extrapancreatic tissues have been of utmost importance; in other words, in the field of diabetes, he has been one of the most world-wide known Swiss investigators. In 1979, Orci succeeded in showing on ultrastructural grounds the localization of insulin biosynthesis in B-cells (11-13). In later studies, he has shown that release of insulin, glucagon and somatostatine occurred by exocytosis, i.e. joining of the inner space of storage secretory granules with the extracellular medium as a result of fusion of the secretory granule's membrane with the cell membrane (14). We also owe to Lelio Orci a great deal of quantitative and biochemical studies concerning the pathways involved in the transformation of hormone precursors into mature hormones (15). His most recent discovery in this field has been the identification of the intracellular site where proteins secreted in response to a specific stimulus are sorted and then released (16, 17). Further important ultrastructural studies of Orci consisted in the demonstration of "non-random" distribution of endocrine cell types within and between islets (18, 19). He demonstrated the presence of specialized areas of cell contact between homologous and heterologous islet cells (20, 21). These studies done in collaboration with his pupils, particularly Alain Perrelet, had a world-wide audience. One of the major discoveries done by Orci was the identification of A-like cells and D-cells in the gastrointestinal tract (22, 23). Using freeze-etching techniques and autoradiography at EM level, his group achieved a pioneer work in the field of ligand binding to cell surface receptors (24).

Recently in Orci's laboratories, in collaboration with R.Montesano and others, methods to study angiogenesis in vitro have been developed. It has been possible to demonstrate that, under appropriate conditions, endothelial cells may invade a reconstituted collagen matrix and organize themselves into capillary-like structures (25-27).

The *Department of Morphology in Berne* is a relatively young institution as far as EM is concerned. Ewald R.Weibel, who is its director since 1966, is a very famous EM investigator. Besides having developed the methodology of morphometry in relation to biological research (see his chapter, as well as (28)), Weibel is an internationally known electron microscopist and lung morphologist.

The early studies achieved in his department on toxicity of 100% oxygen on the lungs (29, 30) constituted a pioneer work in this field. Weibel's main interest in the lung has been to investigate this organ in respect to its gas exchange function, particularly to achieve measurements of the air-blood tissue barrier and estimate its diffusion capacity (31-34). To have a more general approach to oxygen exchange in the body, Weibel together with his co-workers investigated muscle mitochondria (35, 36), and more recently they have extended their studies to the whole respiratory system of different species (37, 38). Although the morphology and morphometry of the lung and gas exchange problems constituted the main research in the Department of Morphology in Berne, other tissues such as the liver (39, 40) have also been investigated.

Another prolific EM study center in Switzerland is the *"Centre de Microscopie Electronique de l'Université de Lausanne"* directed by Alain Gautier. His chapter in this volume gives a large spectrum of the technical contribution of this institution in the development of biological EM research in our country. His overview (41) gives an excellent approach to the problems raised, and solved by EM in the early eighties. One of the most important studies of Gautier's group since 1957 concerned the investigation of thrombocytes (42). These studies, which have been pursued by G.Jean in the early sixties, and later by L.Falcao, have received an international audience (45-47).

A second line of investigations has been done in collaboration with M.Campiche on the lung structures (at present Campiche is professor of general pathology and renal diseases in the Department of Pathology of Lausanne). This investigator has been among the earliest electron microscopists to show the lamellated bodies in type II epithelial cells of the alveolar lining (48). He has later studied the lung of premature infants (49, 50), the ultrastructural morphology of hyalin membrane disease (51), and the modifications of the air-blood barrier in experimentally induced alveolar blood flow changes (52). Campiche has also been one of the earliest EM investigators to study the ultrastructure of alveolar surfactant (53).

Another line of important investigations done in collaboration with Gautier's group was achieved by D.Gardiol in liver morphology and pathology (see pathology). Indeed liver has been one of the interests of the EM Center in Lausanne, and fellows such as F.Minio and others have contributed to these studies (54-57). In the sixties and seventies, kidney has been investigated by O.Bucher and E.Reale (58, 59); the skin, especially pigmented cells of the epidermis by E.Frenk (60-62) and

M.Stoian (63). It should be emphasized that some important work has also been achieved by Gautier and his collaborators in the field of basic research. To cite a few of them, we shall recall H.A.Guénin who has done EM investigations on nuclear structures in the early sixties (64, 65), S.Fakan (66-68) and M.Biggiogera (69) in later periods.

Before closing the chapter on EM investigations done in histology departments, a rapid look to the contribution of Swiss research laboratories on bone ultrastructure has to be made. Charles Baud, professor of histology in the *Department of Morphology in Geneva*, has certainly been the most active contributor in the study of calcified tissue ultrastructure during the last decades (70). His studies on the effect of fluoride (71) and hormones (72) on osteocytic activity have presently obtained international attention. Furthermore he has published in collaboration with René Lagier, from the Department of Pathology of Geneva, some very interesting papers on ochronosis (73). This latter author, together with Baud's pupils, has reported interesting data on the ultrastructure of chondrocalcinosis (74).

Studies performed in the *Departments of Morphology of Zürich and Basel* are reported in other articles published in this volume; the reader is referred to Mühlethaler's and Kellenberger's contributions. We shall however refer to some of them while reviewing the studies accomplished in pathology.

B. Pathology

The *Departments of Pathology* in Switzerland had certainly as much impact as the institutes of histology on the ultrastructural research. In *Zürich*, J.R.Rüttner (75) studied the ultrastructure of ferritine particles in the blood in relation with his investigations on hemochromatosis. Later the study of liver pathology became the main interest in the institutes of pathology, where M.A.Spycher published an important article on the ultrastructural development of the rat embryonal liver (76). Investigations in the *Institute of Histopathology*, the director of which was A.von Albertini, were then directed towards neoplasms and lung pathology, particularly pulmonary dust diseases (77). Parallel to the development of research in liver and lung pathology, EM began to be used in biopsy examination, and quickly it became an unseparable instrument for routine kidney biopsy studies. In 1979, H.Walt started, together with C.Hedinger - co-director of the *Department of Pathology in Zürich* - to investigate testicular germ cells and germ cell tumors. In 1984, he published a very important technical contribution to EM on K4M embedding for immunoelectronics (78). At this period, and up to 1988, the interest in the Department of Pathology of Zürich was particularly focussed on endocrine pathology and the male genital system (79, 80). Furthermore, some contributions to the study of cartilage were published (81), and the systematic study of liver biopsies was started. Moreover EM was used for the rapid diagnosis of viral infections. Then, following the nomination of Theodor Bächi as successor of A.Vogel as the head of the EM center, and of Philipe Heitz as director of the Department of Pathology, a more modern path was given to EM studies.

The two principal articles published recently by these investigators, and which illustrate the scope of EM in Zürich, concern binding and fusion of an enveloped virus (81) and three dimensional views of cells (82).

In the *Department of Pathology of the Basel Medical School*, the motor for EM investigation has been H.Zollinger, and the main research was done on kidney pathology. Nevertheless, prior to the nomination of Zollinger as Head of the Department of Pathology, very interesting contributions were done by F.Gloor on pancreatic vessels (83), and on interstitial cells of the kidney (84). Zollinger´s early studies are summarized in his two review articles (85, 86). A most important pioneer work of Zollinger has been his review on nephrotic syndrome (87), on cryoglobulinaemic nephropathy (88) and Schönlein-Henoch syndrome (89). In fact Zollinger obtained his international celebrity through his studies on analgesic drug induced nephropathy. Together with his pupil and successor, Prof. M.J.Mithatsch, and also Prof. F.J.Gloor - former Director of the Institute of Pathology in St. Gall - (90, 91), he has been one of the first pathologists who insisted on the dangers of phenacetin containing analgesics so frequently used and abused in the past by the Swiss population, particularly by workers in watch factories (92). Later he and his collaborators have shown the very extensive capillary damage caused by phenacetin all through the urinary system (93).

The *electron microscopy in Lausanne* is in some ways synonimous with D.Gardiol and liver. Since 1965, Gardiol has contributed to all important ultrastructural investigations on liver morphology and pathology (56, 57). His most important publications have been related to cholestasis (94-97).

Moreover he extended his EM interest to other tissues such as thyroid (98) and kidney (99).

Besides Gardiol, pathologists such as M.Campiche (formerly on lung ultrastructure, later on kidney biopsy examination), L.Ozzello - former head of the Department of Pathology - and J.Costa - present director of this department - have been largely involved in EM studies, particularly in studying the ultrastructure of breast and soft tissue tumors.

In the *Department of Pathology of Berne*, EM has been more an instrument for routine studies, whereas, in *Geneva*, it occupied a central point of attraction in research and also in routine biopsy investigations.

In the early fifties we have already mentioned investigators such as Rutishauser, Rouiller and Majno who were among the first EM users for the study of bone structures (5). Later, Rouiller, when he became the head of the Institute of Histology, collaborated with young pathologists such as G.Simon and F.Chatelanat in the field of renal pathology (100). However, the explosion of EM coincided with the return of Guido Majno to Geneva as co-chairman of the Department of Pathology. Majno had, during his stay in the USA, published most outstanding articles on inflammation in collaboration with G.E.Palade, who later received the Nobel Price in 1974. Two original articles of these authors are cited in the reference list (101, 102). On the other hand, he had also done a pioneer work with R.S.Cotran on vascular permeability (103). Majno´s most important discovery when he returned to Geneva has been to show the contraction of granulation tissue (104) and of endothelial cells (105). His description of the ultrastructural and functional characteristics of myofibroblast achieved together with G.Gabbiani constituted in 1972 one of the most ambitious works done in

Switzerland (106). In parallel to Gabbiani´s and Majno´s studies on granulation tissue, we reported the presence of contractile cells in pulmonary alveolar interstitium, and named them the contractile interstitial cells of pulmonary alveoli (107). Later we have reported (108) that these cells were identical to the myofibroblasts described by Gabbiani et al. (106, 109, 110). Studies of contractile proteins were pursued in the Department of Pathology after the departure of Majno in 1972. The main group of investigators was that of Gabbiani, but many fellows got interested in this subject. Contractile proteins were shown in aortic endothelium, and their increase was demonstrated in experimental hypertension (111, 112). Later the cytoskeleton of rat aortic smooth muscle cells and those found in intimal thickening were studied (113). The EM studies were conducted together with biochemical and immunohistochemical studies, and thus constituted a basic study for our understanding of mesenchymal cell cytoskeleton (114, 115). Besides myofibroblasts, interstitial cells of the pulmonary alveoli and endothelial cells, EM has been a major investigation instrument for the study of oxygen toxicity on the lungs (116), pulmonary calcinosis (117), modifications of alveolar configuration during respiration (118) and the ultrastructure of some human neoplasms, particularly in respect to the ultrastructure of their cytoskeleton (119-122). Other important contributions have been made in some renal diseases (123), and also in muscle pathology (124-126). Very recently, alveolar tissue changes in experimental GVHR (127) and in bleomycin (128) induced pneumopathies have also been studied by EM. As in all departments of pathology, EM has been used in routine diagnosis work, particularly for kidney biopsies, and for lung and muscle pathology.

Final Comments

This overview concerns particularly the studies done in universtiy departments. Possibly because of the personal concern of the author, more emphasis has been given to pathology than to normal morphology. We would like to express our gratitude to those colleagues who put at our disposal very valuable information concerning the EM research done in their departments. It should however be recalled that the pharmaceutical industry in Switzerland has played a role of utmost importance in the development and explosion of EM research. Investigators such as M.Horisberger from Nestlé, W.Stäubli from Ciba, H.Kuhn, J.G.Richards and J.P.Trauzer from Hoffmann-La Roche, have been pioneers in the EM research program of this country. A special mention has to be addressed to Horisberger who developed the colloidal gold technique, which is at present largely used for EM immunocytochemistry.

Furthermore, EM has been used as an investigation instrument in smaller units in association with major university departments and pharmaceutical industry. In Fribourg for example, Sprumont has accomplished some good studies on ovarian follicules and later on cerebral edema.

Three other institutions where some very outstanding EM research has been done are: 1) ISREC in Lausanne, where H.Isliker, B.Hirst and others have pursued important investigations on cancer and virology research; 2) Biozentrum in Basel; 3) Battelle Institute in Geneva. These different institutions are dealt with in separate contributions included in this book.

A very legitimate question would be to know whether the era of EM is beginning to fade away! Indeed the application of EM to biological research, born in the fifties, grew very rapidly to become a "giant" in the seventies. Then it lost some of its glamour as a result of the discovery of other techniques such as immunohistochemistry, and more recently molecular biology methods. In fact these techniques shed the applications of EM, initially in routine work, later in research. Nevertheless, in 1990, it appears still that EM keeps a predominant role to play in all sorts of morphological investigations of normal or pathological tissues, particularly when it is coupled with other methods. Thus, if the applications of conventional EM appear at present somewhat limited, immunoelectronics, morphometry and computerized image analysis have a tremendous field of investigations ahead.

In concluding, it should be emphasized that the international recognition of many Swiss morphologists occurred thanks to their EM investigations. We owe the world-wide celebrity of E.Weibel to his EM morphometric studies on the lung; our international rank in research on diabetes to L.Orci´s ultrastructural investigations; our knowledge in ultrastructure of cholestasis to D.Gardiol; the initiation of cancer research to A.Vogel and A.von Albertini; the study of pulmonary dust disease to J.Rüttner, and our knowledge of myofibroblasts and endothelial cells as well as of their cytoskeleton to G.Gabbiani s work. Last but not least, we owe to E.Kellenberger the most important advancement in bacteriophage studies through which the new era of molecular biology was born.

Acknowledgements

We thank Mrs. Claire-Lise de Marignac and Miss Alberte Polichouk for their valuable technical assistance.

References

1. E.Kellenberger and W.Chiu, Ultramicroscopy **10**, 165 (1982).
2. C.Oberling, W.Bernhard, L.Febure and J.Hamel, Comptes-Rendus du 1er Congrès International de Microscopie Electronique, Paris, 14-22 Sept. 1950. Ed. par Revue d'Optique Théorique et Instrumentale, Mémoires hors série **1**, 600 (1953).
3. A.Vogel, 4.Tagung für Elektronenmikroskopie des Schweiz.Komitees für Optik, Zürich, 23.5.1957.
4. A.Vogel and U.W.Schnyder, Dermatologica **135**, 149 (1967).
5. E.Rutishauser, L.Huber, E.Kellenberger, G.Majno and C.Rouiller, Arch.des Sci.Soc.Phys.& Hist.naturelle de Genève **3**, 175 (1950).
6. C.Rouiller and A.M.Jézéquel, Electron Microscopy of the Liver, in: The Liver, Morphology, Biochemistry, Physiology, **1**, ed.C.Rouiller, Academic Press, New York, 1963, p. 195.
7. P.Nicolescu and C.Rouiller, Z.Zellforsch. **76**, 313 (1967).
8. C.Rouiller, R.Pictet, P.Nicolescu and L.Orci, Rev.Int.Hépatologie **17**, 827 (1967).
9. A.Matter, L.Orci, W.G.Forssmann and C.Rouiller, J.Ultrastruct.Res. **23**, 272 (1968).
10. A.Matter, L.Orci and C.Rouiller, J.Ultrastruct.Res. Suppl. **11**, 1 (1969).
11. L.Orci, A.E.Lambert, Y.Kanazawa, M.Amherdt, C.Rouiller and A.E.Renold, J.Cell Biol. **50**, 565 (1971).
12. L.Orci, K.H.Gabbay and W.J.Malaisse, Science **175**, 1128 (1972).
13. L.Orci, A.A.Like, M.Amherdt, B.Blondel, Y.Kanazawa, E.B.Marliss, A.E. Lambert, C.B.Wollheim and A.E.Renold, J.Ultrastruct.Res. **43**, 270 (1973).

14. L.Orci, A.Perrelet and D.S.Friend, J.Cell Biol. **75**, 23 (1977).
15. R.G.W.Anderson and L.Orci, J.Cell Biol. **106**, 539 (1988).
16. L.Orci, M.Ravazzola, M.Amherdt, A.Perrelet, S.K.Powell, D.L.Quinn and H.P.H.Moore, Cell **51**, 1039 (1987).
17. J.E.Rothman and L.Orci, FASEB J. **4**, 1460 (1990).
18. L.Orci, F.Malaisse-Lagae, D.Baetens and A.Perrelet, Lancet **2**, 1200 (1978).
19. F.Malaisse-Lagae, Y.Stefan, J.Cox, A.Perrelet and L.Orci, Diabetologia **17**, 361 (1979).
20. P.Meda, A.Perrelet and L.Orci, Modern Cell Biol. **3**, 131 (1984).
21. D.Bosco, L.Orci and P.Meda, Exp.Cell Res. **184**, 72 (1989).
22. L.Orci, R.Pictet, W.G.Forssmann, A.E.Renold and C.Rouiller, Diabetologia **4**, 56 (1968).
23. W.G.Forssmann, L.Orci, R.Pictet, A.E.Renold and C.Rouiller, J.Cell Biol. **40**, 692 (1969).
24. J.L.Carpentier, P.Gorden, M.Amherdt, E.van Obberghen, C.R.Kahn and L.Orci, J.Clin.Invest. **61**, 1057 (1978).
25. J.L.Carpentier, P.Gorden, P.Freychet, A.Le Cam and L.Orci, J.Clin. Invest. **63**, 1249 (1979).
26. R.Montesano, P.Mouron, M.Amherdt and L.Orci, J.Cell Biol. **97**, 935 (1983).
27. R.Montesano, L.Orci and P.Vassalli, J.Cell Biol. **97**, 1648 (1983).
28. E.R.Weibel, Histochem.Cytochem. **29,** 1043 (1981).
29. G.S.Kistler, P.R.B.Caldwell and E.R.Weibel, J.Cell Biol. **32**, 605 (1967).
30. Y.Kapanci, E.R.Weibel, H.P.Kaplan and R.F.Robinson, Lab.Invest. **20**, 101 (1969).
31. E.R.Weibel and J.Gil, Respir.Physiol. **4**, 42 (1968).
32. E.R.Weibel, Respir.Physiol. **11**, 54 (1970/71).
33. P.Gehr, M.Bachofen and E.R.Weibel, Respir.Physiol. **32**, 121 (1978).
34. E.R.Weibel, L.B.Marques, M.Constantinopol, F.Doffey, P.Gehr and C.R.Taylor, Respir.Physiol. **69**, 81 (1987).
35. H.Hoppeler, P.Lüthi, H.Claassen, E.R.Weibel and H.Howald, Pflügers Arch. **344**, 271 (1973).
36. H.Hoppeler, H.Howald, K.Conley, S.L.Lindstedt, H.Claassen and E.R. Weibel, J.Appl.Physiol. **59**, 320 (1985).
37. E.R.Weibel and C.R.Taylor, Respir.Physiol. **44**, 1 (1981).

186

38. C.R.Taylor, R.H.Karas, E.R.Weibel and H.Hoppeler, Respir.Physiol. **69**, 1 (1987).
39. E.R.Weibel, W.Stäubli, H.R.Gnägi and F.A.Hess, J.Cell Biol. **42**, 68 (1969).
40. R.P.Bolender, D.Baumgartner, G.Losa, D.Muellener and E.R.Weibel, J.Cell Biol. **77**, 565 (1978).
41. A.Gautier, Microscopie électronique conventionelle en biologie: comment suivre un fil d´Ariane dans un labyrinthe si complexe?, in: Impact de la microscopie électronique sur les sciences expérimentales, ed. P.A.Buffat and T.Jalanti, AVCP, Lausanne 1980, p. 335.
42. R.Feissly, A.Gautier and I.Marcovici, Rev.Hémat. **12**, 397 (1957).
43. G.Jean, L.Racine, R.Marx and A.Gautier, Thrombos.Diathes. Haemorrh. **9**, 1 (1963).
44. A.Gautier, G.Jean, M.Probst and L.Falcao, Arch.Ital.Anat.Istol.Pat. **37**, 503 (1963).
45. L.Falcao and A.Gautier, Blut **16**, 57 (1967).
46. L.Falcao, Coagulation **1**, 229 (1968).
47. L.Falcao, Blut **28**, 337 (1974).
48. M.Campiche, J.Ultrastruct.Res. **3**, 302 (1960).
49. M.Campiche, S.Prod´hom and A.Gautier, Ann.Pédiat. **196**, 81 (1961).
50. M.Campiche, A.Gautier, E.I.Hernandez and A.Reymond, Pediatrics **32**, 976 (1963).
51. M.Campiche, M.Jaccottet and E.Juillard, Ann.Pédiat. **199**, 74 (1962).
52. A.Gautier, M.Campiche and E.I.Hernandez, Mikroskopie **19**, 54 (1964).
53. M.Campiche, Proc.Internat.Union Physiol.Sci. **6**, 131 (1968).
54. F.Minio, A.Gautier and P.Magnenat, Z.Zellforsch. **72**, 168 (1966).
55. F.Minio, L.Lombardi and A.Gautier, J.Ultrastruct.Res. **16**, 339 (1966).
56. F.Minio, P.Magnenat, D.Gardiol and A.Gautier, Z.Zellforsch. **65**, 47 (1965).
57. F.Minio, G.Alberti, D.Gardiol, A.Gautier, P.Magnenat and A.Torsoli, Z.Zellforsch. **66**, 496 (1965).
58. O.Bucher and E.Reale, Z.Zellforsch. **54**, 167 1961).
59. E.Reale, V.Marinozzi and O.Bucher, Acta Anat. **52**, 22 (1963).
60. E.Frenk, Arch.klin.exp.Derm. **235**, 16 (1969).
61. E.Frenk, Dermatologica **143**, 12 (1971).

62. E.Frenk, Dermatologica **159**, 185 (1979).
63. M.Stoian, Dermatologica **141**, 95 (1970).
64. H.A.Guénin and A.Gautier, Rev.Suisse Zool. **67**, 210 (1960).
65. H.A.Guénin, Mikroskopie **19**, 54 (1964).
66. S.Fakan, G.Leser and T.E.Martin, J.Cell Biol. **98**, 358 (1984).
67. S.Fakan, Methods Achiev.Exp.Pathol. **12**, 105 (1986).
68. S.Fakan, G.Leser and T.E.Martin, J.Cell Biol. **103**, 1153 (1986).
69. M.Biggiogera, Basic Appl.Histochem. **30**, 501 (1986).
70. C.A.Baud, Clin.Orthop.Rel.Res. **56**, 227 (1968).
71. C.A.Baud and G.Boivin, Metab.Bone Dis.Rel.Res. **1**, 49 (1978).
72. C.A.Baud and G.Boivin, Clin.Orthop. **136**, 270 (1978).
73. R.Lagier, C.A.Baud, D.Lacotte and T.Cunningham, Am.J.Clin.Pathol. **90**, 95 (1988).
74. G.Boivin and R.Lagier, Virchows Arch.(Pathol.Anat.) **400**, 13 (1983).
75. J.R.Rüttner, M.A.Spycher and H.E.Brunner, Med.Exp. **7**, 379 (1962).
76. M.A.Spycher, Path.Mikrobiol. **30**, 303 (1967).
77. H.Sticher, M.A.Spycher and J.R.Rüttner, Nature **241**, 49 (1973).
78. H.Walt and B.Armbruster, Cell Tissue Res. **236**, 487 (1984).
79. M.A.Spycher and U.N.Wiesmann, Verh.Dtsch.Ges.Path. **66**, 203(1982).
80. A.Martin, C.Hedinger, M.Häberlin and H.Walt, Virchows Arch.B Cell. Pathol. **55**, 159 (1988).
81. T.Bächi, J.Cell Biol. **107**, 1689 (1988).
82. C.Bron, P.Grémillet, D.Launay, M.Jourlin, H.P.Gautschi, T.Bächi and J.Schüpbach, J.Microsc. **157**, 115 (1989).
83. F.Gloor, Acta Anat. **35**, 63 (1958).
84. F.Gloor and L.A.Neiditsch-Halff, Z.Zellforsch. **66**, 488 (1965).
85. H.U.Zollinger, Pathol.Europ. **5**, 2 (1970).
86. H.U.Zollinger, J.Moppert, G.Thiel and H.P.Rohr, Curr.Top.Pathol. **57**, 1 (1973).
87. H.U.Zollinger, Postgrad.Med.J. **45**, 701 (1969).
88. A.Tarantino, A.De Vecchi, G.Montagnino, E,Imbasciati,M.J.Mihatsch, H.U.Zollinger, G.Bargiano di Belgiojoso, G.Busnach and C.Ponticelli, Q.J.Med. **197**, 1< (1981).
89. H.U.Zollinger,M.J.Mihatsch,F.Gaboardi,G.Banfi,A.Edefonti,M.Bardare and F.Gudat, Virchows Arch.A Path.Anat.Histol. **388**, 155 (1980).
90. F.J.Gloor, Ergebn.allg.Path.path.Anat. **41**, 64 (1961).

91. F.J.Gloor, Some morphologic features of chronic interstitial nephritis (chronic pyelonephritis) in patients with analgesic abuse; in: Progr. in Pyelonephritis, 1965, p. 287.

92. M.J.Mihatsch, J.Torhorst, E.Steinmann, H.Hofer, M.Stickelberger, L. Bianchi, K.Berneis and H.U.Zollinger, Path.Res.Pract. **164**, 68 (1979).

93. M.J.Mihatsch, H.O.Hofer, F.Gudat, C.Knüsli, J.Torhorst and H.U. Zollinger, Clin.Nephrol. **20**, 285 (1983).

94. R.Picardi, D.Gardiol and A.Gautier, Z.Zellforsch. **84**, 311 (1968).

95. R.Picardi, D.Gardiol and A.Gautier, Z.Zellforsch. **84**, 319 (1968).

96. F.Bonvicini, A.Gautier, D.Gardiol and G.A.Borel, Lab.Invest. **38**, 487 (1978).

97. H.Loosli, D.Gardiol and A.Gautier, Virchows Arch.(Cell Pathol.) **35**, 213 (1981).

98. M.Benathan, T.Lemarchand-Béraud, A.Gautier and D.Gardial, Virchows Arch.(Cell Pathol.) **44**, 323 (1983).

99. S.Gomba, D.Gardiol and A.Gautier, Path.Europ. **8**, 61 (1973).

100. G.Simon and F.Chatelanat, Ultrastructure of the normal and pathological glomerulus; in: The Kidney. Morphology, Biochemistry and Physiology, **1**, ed. C.Rouiller and A.Muller, Academic Press, New York (1967).

101. G.Majno and G.E.Palade, J.Biophys.Biochem.Cytol. **11**, 571 (1961).

102. G.Majno, G.E.Palade and G.I.Schoefl, J.Biophys.Biochem.Cytol. **11**, 607 (1961).

103. R.S.Cotran and G.Majno, Protoplasma **43**, 45 (1967).

104. G.Majno, G.Gabbiani, B.J.Hirschel, G.B.Ryan and P.R.Statkov, Science **173**, 548 (1971).

105. G.Majno, S.M.Shea and M.Leventhal, J.Cell Biol. **22**, 227 (1969).

106. G.Gabbiani, B.J.Hirschel, G.B.Ryan, P.R.Statkov and G.Majno, J.Exp.Med. **135**, 719 (1972).

107. Y.Kapanci, A.Assimacopoulos, C.Irlé, A.Zwahlen and G.Gabbiani, J.Cell Biol. **60**, 375 (1974).

108. Y.Kapanci, P.Mo Costabella, P.Cerutti and A.Assimacopoulos, Distribution and function of cytoskeletal proteins in lung cells with particular reference to "contractile interstitial cells"; in: Methods and Achievements in Experimental Pathology, ed. G.Jasmin and M.Cantin, Karger, Basel, 1979, p.147.

109. G.Gabbiani, G.B.Ryan and G.Majno, Experientia **27**, 549 (1971).

110. E.Rüngger-Brandle and G.Gabbiani, Am.J.Pathol. **110**, 361 (1983).

111. G.Gabbiani, M.C.Badonnel and G.Rona, Lab.Invest. **32**, 227 (1965).

112. G.Gabbiani, G.Elemer, C.Guelpa, M.B.Vallotton, M.C.Badonnel and I.Hüttner, Am.J.Pathol. **96**, 399 (1979).

113. O.Kocher, O.Skalli, W.S.Bloom and G.Gabbiani, Lab.Invest. **50**, 645 (1984).

114. O.Skalli and G.Gabbiani, The biology of the myofibroblast: relationship to wound contraction and fibrocontractive diseases; in: The Molecular and Cellular Biology of Wound Repair, chap. 17, ed. R.A.F.Clark and P.M.Henson, Plenum Publ. Corp., New York, 1988, p. 373.

115. O.Skalli, W.Schürch, T.Seemayer, R.Lagacé, D.Montandon, B.Pittet and G.Gabbiani, Lab.Invest. **60**, 275 (1989).

116. V.E.Gould, R.Tosco, R.F.Wheelis, N.S.Gould and Y.Kapanci, Lab.Invest. **26**, 499 (1972).

117. J.Eggermann and Y.Kapanci, Lab.Invest. **24**, 469 (1971).

118. A.Assimacopoulos, R.Guggenheim and Y.Kapanci, Lab.Invest. **34**, 10 (1976).

119. Y.S.Fu, G.Gabbiani, G.I.Kaye and R.Lattes, Cancer **35**, 176 (1975).

120. G.Gabbiani, Y.S.Fu, G.I.Kaye, R.Lattes and G.Majno, Cancer **30**, 486 (1972).

121. G.Gabbiani, G.I.Kaye, R.Lattes and G.Majno, Cancer **28**, 1031 (1971).

122. W.Schürch, O.Skalli, R.Lagacé, T.A.Seemayer and G.Gabbiani, Am.J. Pathol. **136**, 771 (1990).

123. G.Gabbiani, M.C.Badonnel and P.Vassalli, Lab.Invest. **32**, 33 (1975).

124. P.Hoffmeyer, J.N.Cox, Y.Blanc, J.M.Meyer and W.Taillard, Clin.Orthop. Rel.Res. **232**, 112 (1988).

125. P.Hoffmeyer, J.N.Cox and D.Fritschy, Int.Orthop. **11**, 53 (1987).

126. P.Hoffmeyer, C.Freuler and J.N.Cox, in press in Int.Orthop. (1990).

127. P.F.Piguet, G.E.Grau, M.A.Collart, P.Vassalli and Y.Kapanci, Lab.Invest. **61**, 37 (1989).

128. P.F.Piguet, M.Collart, G.E.Grau, Y.Kapanci and P.Vassalli, J.Exp.Med. **170**, 655 (1989).

CRYOTECHNIQUES AND RELATED METHODS

Hans Moor
Institute of Cell Biology
Federal Institute of Technology
Zürich

Since the end of the 19th century, biologists have used freezing techniques for stabilization of cells and tissues in order to overcome the disadvantages of chemical fixation and embedding. E.g. frozen sections have been used extensively for light microscopic histochemistry. With the introduction of the electron microscope severe problems arose concerning specimen preparation caused by the additional requirements of the new tool. In the endeavour to omit chemical and structural alterations resulting from the preparation for conventional ultrathin sectioning, H.R.Müller (1) in 1957 made first steps in introducing freezing techniques. He has snap-frozen tiny pieces of leaf by dipping in LN_2-cooled propane; then he subjected them to freeze-drying at -60°C; subsequently the specimens were vacuum-infiltrated with metacrylate and UV-polymerized at -10°C.

192

This was the best approach possible in those days, but it failed because the temperature during drying was too high and because of deleterious interactions between embedding medium and the chemically and physically non stabilized specimen.

Discussions with H.R.Müller and Russell Steere, who visited the Laboratory of Electron Microscopy in Zurich in 1957, showed that a completely alternative method should be applied, the so-called "freeze-etching" (2): Snap-frozen samples are cleaved at about -100°C, "etched" by superficial freeze-drying and the fracture faces coated with heavy metal and carbon. After thawing the specimen, the metal carbon replica is stripped off, cleaned and then viewed in the microscope. This methodological idea was the starting point for a whole series of cryotechnical developments, performed in the former Institute of General Botany and the later Institute of Cell Biology. The work was promoted first by A.Frey-Wyssling, and then by K.Mühlethaler and H.Moor.

First, an apparatus had to be constructed which guaranteed for reproducible freeze-etching (3). Then, a suitable specimen had to be investigated in order to show the real cytological value of the method (4). Subsequently, it became evident that the cryofixation technique had to be improved (5) and also the freeze-fracturing, that is the vacuum- and the evaporation-technique (6). Electron beam heated evaporators were developed and applied for shadowing specimens cooled down to LN$_2$ temperature under ultrahigh vacuum conditions (7) in order to improve the resolution power of the replicas. Further improvements have been achieved by reducing the temperature to -260°C (8). According to these technical developments, in 1964 the first high-vacuum freeze-etcher and in

1982 the first ultrahigh-vacuum freeze-fracturing apparatus were produced on a commercial basis.

The very thin, high-resolving and incoherent shadowing films became so "noisy" that the results had to be quantified by computer aided image averaging (8). In order to facilitate the interpretation of details of the structural record, surface relief reconstruction based on averaged data and computer aided simulation of the shadowing process have been undertaken (9). Our knowledge about such films was a great help for the first steps in adopting biological macromolecules to the claims of the scanning tunnelling microscope (10).

Under construction is now a small issue of an UHV preparation chamber which shall be connected to a transmission electron microscope in order to enable the direct transfer of freeze-dried and shadowed specimens into the column, that is without contact with the atmosphere. The on-line preparation concept will also be an essential tool for high resolution scanning electron and tunnelling microscopy.

Molecular specimens, suitable for highest resolution microscopy do not require a very elaborated freezing technique; such moist specimens, adsorbed at a carbon film are so thin that they can be snap-frozen by dipping in LN_2 at -210°C. Essential problems arise from much thicker samples like tissue slices. Very high freezing rates are required, if the ice crystal size shall be reduced so far that it does not disturb the structural record. Many techniques have been developed, among them the propane-jet freezing (11, 12), which enable specimens to be vitrified in a thickness of up to 10 µm. The low thermal conductivity of water prevents the maintaining of sufficiently high cooling rates in thicker specimens, unless its freezing properties are altered. One way of achieving this goal is the introduction of an antifreeze agent (5); an alternative method, which is structurally and physiologically much less harmful, consists in freezing under high hydrostatic pressure (13). Many conceptual and technological problems had to be overcome until this technique became a laboratory routine (14) and an apparatus has been developed for commercial purposes (15).

This technique has offered the opportunity for a new approach to freeze-substitution (16). From fresh tissue slices through pressure freezing, substitution in methanol starting at -90°C and low temperature embedding in LOWICRYL, the production of ultrathin sections containing well preserved cytochemical information has been attained (17). In addition, we are on the way to develop an on-line freeze-fracture technique: the already mentioned UHV preparation unit will also enable the direct transfer of freeze-fractured and coated specimens into a high resolution scanning microscope.

References

1. H.R.Müller, J.Ultrastruct.Research **1**, 109 (1957).
2. R.L.Steere, J.Biophys.Biochem.Cytol. **3**, 45 (1957).
3. H.Moor, K.Mühlethaler, H.Waldner and A.Frey-Wyssling, J.Biophys.Biochem.Cytol. **10**, 1 (1961).
4. H.Moor and K.Mühlethaler, J.Cell.Biol. 17, 609 (1963).
5. H.Moor, Z.Zellforsch. **62**, 546 (1964).
6. H.Moor, Philos.Trans.Royal Soc.London, Ser.B,**1971**, 121.
7. H.Gross, E.Bas and H.Moor, J.Cell.Biol. **76**, 712 (1978).
8. H.Gross, T.Müller, I.Wildhaber and H.Winkler, Ultramicrosc. **16**, 287 (1985).
9. H.Winkler and H.Gross, Scanning Microsc. Suppl.2, 379 (1988).
10. M.Amrein, A.Stasiak, H.Gross and G.Travaglini, Science **240**, 514 (1988).
11. H.Moor, J.Kistler and M.Müller, Experientia 32, **805** (1976).
12. M.Müller, T.Marti and S.Kriz, Electron Microsc. 1980, **2,**Proc.7th Europ.Congr.Electron Microsc., Leiden, 720.
13. H.Moor and U.Riehle, Electron Microsc. 1968, **2,**Proc. 4th Europ. Reg.Conf. Electron Microsc., Rome, 33.
14. H.Moor, G.Bellin, C.Sandri and K.Akert, Cell Tissue Res. **209**, 201 (1980).
15. H.Moor, Electron Microsc. 1986, **3**, Proc. 11th Int.Congr.Electron Microsc., Kyoto, 1961.
16. M.Müller and H.Moor, Sci. Biol.Preparation, SEM Inc., AMF O´Hare, Chicago, 131 (1984).
17. R.A.Steinbrecht and M.Müller, in: Cryotechniques in Biological Electron Microscopy (R.A.Steinbrecht, K.Zierold eds.), 149, Springer Berlin (1987).

History of Electron Microscopy
in Switzerland
Edited by John R. Günter
© 1990 Birkhäuser Verlag Basel

THE CONTRIBUTIONS OF SWITZERLAND TO THE DEVELOPMENT OF MORPHOMETRY AND STEREOLOGY

Ewald R.Weibel
Department of Anatomy
University of Berne

Quantitative methods were introduced relatively late into microscopy. One reason was that measurements obtained on the flat images presented by sections were of limited and questionable significance in terms of the three dimensional structure. This problem became accentuated in biological electron microscopy with the use of ultrathin sections of cells and tissues, and even more so in the study of polished sections of materials such as metals and minerals. On the other hand, purely descriptive microscopic investigations could only incompletely satisfy the need for structural information. The increasing sophistication of scientific questions asked pressed for the development of suitable methods for deriving true three-dimensional morphometric data from measurements obtained on flat images of random sections. This need was clearly felt in the early 1960's, particularly when one became aware that the electron microscopic image of cells was quite monotonous, with repetitive basic patterns of structure modulated mostly quantitatively from one cell type to another.

Up to that time some simple methods of microscopic measurement had been developed pragmatically by practitioners of microscopy in various disparate fields, methods which largely lacked a solid theoretical foundation. In 1963 the International Society for Stereology (ISS) was founded. It first brought together microscopists from biology and materials sciences, with only a few mathematicians joining. Its goal was to advance the development of microscopic measurement on sections, and it succeeded in this largely through intense interdisciplinary collaboration.

The Society had a very strong European base, and the Secretariat was located in Switzerland, with E.R.Weibel serving as the first Secretary, first in Zürich and then in Berne; the Secretariat returned to Berne in 1983-1987 with Luis Cruz-Orive as Secretary. In 1967 E.R.Weibel became the second President of the ISS and organized the 3rd International Congress for Stereology in Berne in 1971, with the strong and active support by the SGOEM that had taken a significant interest in these new methods. When the ISS joined the Royal Microscopical Society (London) in editing the Journal of Microscopy and declared it its official publication the Editorial Office for Stereology of the Journal of Microscopy was set up in Berne from 1971-1978.

In the development of stereology the congress of 1971 was to some extent a turning point, as a special effort was made to bring in mathematicians and to stimulate them to develop sound theoretical foundations for stereological methods, particularly with respect to sampling problems. That this trend was initiated in Berne was not fortuitous: one of the true mathematical foundations for stereology, namely integral geometry, had been laid in Berne by the late professor of mathematics Hugo Hadwiger (1908-1981). Although Hadwiger remained in the field of "pure mathematics" some of his disciples, mainly Hans Giger, undertook to derive theoretical foundations for practical stereological methods which proved to be a great stimulus for further advances. These advances then took on a rapid pace at an international level, notably by the work of Roger E.Miles in Australia who had spent some time in Berne and retained close relations with the Department of Anatomy. Here a group working on theoretical stereology was built up with Luis Cruz-Orive as the leader, whose main focus was to develop practical methods for microscopic measurement based on a sound theory of stereology. These efforts led to significant advances in the degree of sophistication of stereological methods, such as new and more efficient sampling strategies, corrections for biases due to section thickness or to microscopic resolution (the latter based on concepts of fractal geometry), or special methods for dealing with anisotropic structures. In recent years, the group in Berne has significantly contributed to the development of a new generation of stereological methods, partly using paired rather than single sections, which eliminate many of the limitations adhering to the more "classical" methods. The practical application of these methods requires micrographs to be "sampled" with suitable, geometrically defined test systems ("grids") composed of points, lines, and areas (Fig.1). The group in Berne developed an extensive set of such test grids and defined the rules for their application

Fig. 1a)

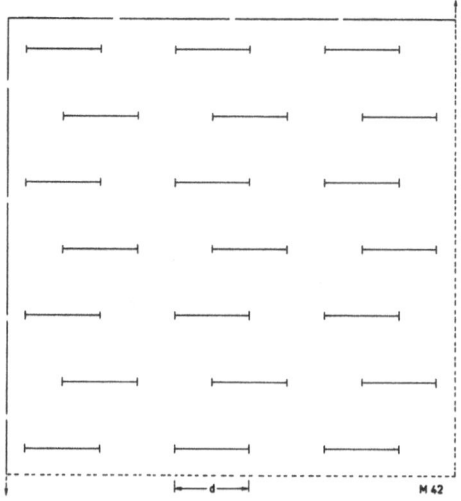

Fig.1b)

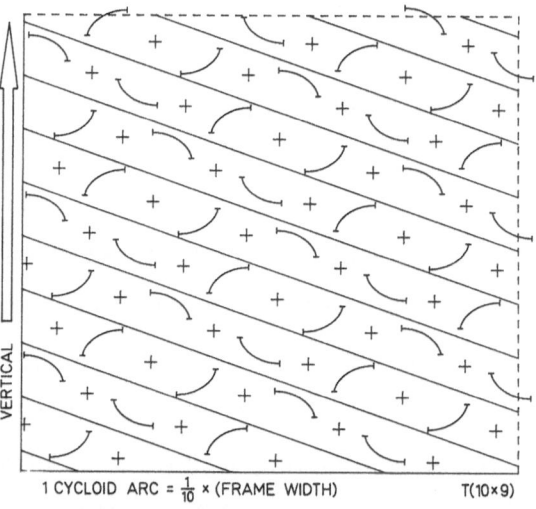

Fig.1: Test systems for stereology called coherent because they comprise test points, lines and area for estimating volume, surface and number densities.
a) So called multipurpose test system from 1966 (1);
b) advanced cycloid test system for anisotropic structures (2).

(see (3)). In addition, computer programs were developed that allowed work to be done efficiently and reliably, in keeping with the evolution of computers.

While stereological methods can often be applied in practical microscopy with very little technology besides some test grids and a reasonably efficient data handling system (for which a small personal computer is usually adequate), some steps are made more efficient by specialized instruments. Through collaboration with WILD Heerbrugg a sampling stage light microscope was developed in the 1960´s (Fig.2), as well as a set of useful test grids. Furthermore, the Swiss firm KONTRON was among the first to produce a semiautomatic tracing device that could be used for stereological work; from there more elaborate image analyzing devices were subsequently developed. In spite of their high level of technological sophistication their usefulness for stereological morphometry remains limited, however.

Fig.2: Automatic sampling stage microscope developed by WILD Heerbrugg in 1964.

In addition, a number of Swiss groups were highly active in developing practical morphometric methods which they applied in their research, particularly in biology and medicine. Some fields of application were the analysis of bone structure by R.K.Schenk in Basel and later in Berne, the electron microscopic study of liver cells by H.P.Rohr in Basel, and the morphometric study of lung, muscle and liver cells by the group of E.R.Weibel in Berne. Much of this work proceeded in parallel with the theoretical and methodological developments described above. The combination of theoretical development of methods with their use in solving problems of structural analysis by means of electron microscopy was largely responsible for the rapid and significant advances of stereology as the fundamental approach for morphometry.

References

1. E.R.Weibel, G.S.Kistler and W.F.Scherle, J.Cell Biol. **30,** 23 (1966).
2. A.J.Baddeley, H.J.G.Gundersen and L.M.Cruz-Orive, J.Microsc. **142,** 259 (1986),
   L.M.Cruz-Orive, Acta Stereolog. **6/III,** 3 (1987).
3. E.R.Weibel, Stereological Methods, Vol. 1+2, Academic Press, 1979/80.

V. SWISS SOCIETY FOR OPTICS AND ELECTRON MICROSCOPY

HISTORY OF THE SWISS SOCIETY
FOR OPTICS AND ELECTRON MICROSCOPY
(SGOEM / SSOME)

John R. Günter
Institute for Inorganic Chemistry
University of Zürich

Electron microscopy began to be established in Switzerland during and immediately after the end of world war II (see chapters on early times of electron mircrcoscopy in Switzerland). As early as in 1946, there were already two instruments of the Swiss manufacturer Trüb, Täuber & Cie.AG in operation, at the Institutes for Chemistry in Berne and at the Institute for Physics in Geneva. The number of instruments and laboratories involved in electron microscopy continued to grow steadily, though slowly (Table 1). The early electron microscopists had of course more or less loose personal contacts, but the formation of international scientific unions and of

UNESCO after the end of the war called for more formal relations in order to be represented in these organizations.

This became obvious for the first time, in the field of optics, at a meeting in Paris in October 1946, when international cooperation was discussed. As a consequence, the Internation Union of Pure and Applied Physics (IUPAP) founded in January 1947 the "Commission Internationale d'Optique - CIO", which was affiliated to IUPAP and to UNESCO. A group of "18 scientifically-technically oriented opticians" (citation from the records of their meeting) in Switzerland felt the need of participating in the newly founded CIO, and for this purpose formed, together with 12 other members, who were not personally attending the meeting, a provisional committee in Zürich on April 23, 1948, called "Schweizerisches Komitee für Optik / Comité Suisse d'Optique - SKO/CSO" (Swiss Committee for Optics) as a subcommission of CIO, and affiliated in Switzerland to the "Comité Suisse de Physique". Its first president was H.König, also president of the already existing "Schweizerisches Beleuchtungskomitee" (Swiss Committee for Lighting), its first secretary was W.Lotmar. The constitution of SKO/CSO was worked out slowly and became valid only in 1949 (thus to be considered as the actual year of its foundation), just in time for sending official delegates to the International Optical Conference in London in 1950. It is noteworthy that this constitution already mentions an interest of the SKO/CSO in electron optics. (It may also be mentioned that the membership fee at that time was a modest SFr. 5.-- for personal members and Sfr. 50.-- for cooperate members; and that is has since been increased only once, in 1976, to Sfr. 10.-- for personal members).

After a change in the presidency, with N.Schätti becoming the new head of SKO/CSO, the late professor A.Frey-Wyssling († 1988) from Zürich suggested in 1954 at a meeting devoted entirely to lectures on electron microscopy a change of the name and constitution, in order to show more clearly that electron microscopy had become a major concern of SKO/CSO. The new constitution was installed in 1955 and the committee renamed to "Schweizerisches Komitee für Licht- und Elektronenopik - SKO/CSO". The secretariat was divided into sections for Optics and for Electron Microscopy, the first secretary for electron microscopy being A.Gautier from Lausanne. (The further succession of presidents and secretaries of the electron microscopy and optics sections is shown in table 2). Under its new form, the newly organized SKO/CSO joined IFSEM (then called IFEMS) at its foundation in 1955. It became affiliated to the Swiss Academy of Natural Sciences (Schweizerische Naturforschende Gesellschaft, SNG) in 1963, still as a subcommittee of the Comité Suisse de Physique.

The quickly increasing membership (Table 3) soon made this form of a subcommittee inadequate, and a further revision of the constitution was induced, aiming at becoming a scientific society in its own rights. These efforts were successful in 1969, when the new constitution was accepted by the Swiss Academy of Natural Sciences (then SNG, now SANW), and the newly named "Schweizerische Gesellschaft für Optik und Elektronenmikroskopie / Société Suisse d'Optique et de Microscopie Electronique - SGOEM / SSOME" (Swiss Society for Optics and Electron Microscopy) was admitted as a full member to the Academy. Partial revisions of the constitution followed in 1975 and 1978. In 1976 the Society also became a member of CESEM, when this body was founded at the 6th European Congress on Electron Microscopy in Jerusalem. Members of the Swiss society were elected twice as chairmen of CESEM since (E.Kellenberger and J.R.Günter). In 1987, SGOEM / SSOME also became a member of the Swiss Academy of Technical Sciences, SATW, as well as of EUROPTICA.

The SGOEM/SSOME edits a quarterly bulletin for information of its members about current affairs, new books and coming events, and helds regular meetings (the locations and dates of past meetings are listed in table 4). In the recent years, it has also started to organize specialized courses and workshops in the field of electron microscopy.

The following members of the Society have been elected as honorary members: N.Schätti, L.Wegmann (Fig. 1), E.Kellenberger, A.Gautier and W.Villiger.

Fig. 1:   The late president Lienhard Wegmann (right) in discussion with Prof.H. Bethge (GDR) at the SGOEM/SSOME meeting in Vaduz, 1982. (Photograph by R.Guggenheim).

TABLE 1:

Number of Transmission Electron Microscopes in Switzerland

(according to the records of the Society)

1946	2 tems		(Univ. Berne, Univ. Geneva)
1947	2 tems,	1 diffractograph	(+ LSRH Neuchâtel)
1948	3	1	(+ ETH Zürich)
1951	4	1	(+ Univ. Geneva)
1952	6	2	(+ Univ. Geneva, Univ.Berne, ETH Zürich)
1954	8	4	(+ BBC Baden, Battelle Geneva, CME Univ. Lausanne)
1955	9	4	(+ Inst.Straumann, Waldenburg)
1956	10	5	(+ Inst.Straumann, Waldenburg, Univ. Berne)
1957	14	5	(+ CME Lausanne, Univ. Basel, ETH Zürich, Alusuisse AG Neuhausen)
1958	16	5	(+ BBC Baden, LSRH Neuchâtel)

1958:   16 transmission electron microscopes,
          5 electron diffractographs, in 10 laboratories

1964:   40 transmission electron microscopes, in 28 laboratories.

Number of Scanning Electron Microscopes in Switzerland

(according to the lists published annually by the Society)

1974	20	1983	99
1976	39	1984	106
1977	44	1985	115
1978	59	1986	124
1979	66	1987	136
1980	76	1988	153
1981	84	1989	153
1982	92	1990	157

TABLE 2:

Presidents and Secretaries

Schweizerisches Komitee für Optik / Comité Suisse d'Optique
SKO / CSO

	President	Secretary
	President	Secretary
1947-1952	H.König	W.Lotmar
1953-1954	N.Schätti	W.Lotmar

Schweizerisches Komitee für Licht- und Elektronenoptik
Comité Suisse d'Optique Photonique et Electronique
SKO / CSO

	President	Secretary El.Micr.	Secretary Optics	
1955-1960	N.Schätti	A.Gautier	W.Lotmar	(55-56)
			E.Millet	(57-62)
1961-1966	N.Schätti	L.Wegmann	A.Werfeli	(63-66)
1967-1968	L.Wegmann	M.Gribi	R.David	(67-68)

Schweizerische Gesellschaft für Optik und Elektronenmikroskopie
Société Suisse d'Optique et de Microscopie Electronique
SGOEM / SSOME

1969-1970	L.Wegmann	M.Gribi	R.David	(69-70)
1971-1972	L.Wegmann	W.Stäubli	C.v.Planta	(71-72)
1973-1976	L.Wegmann	G.Kistler	F.K.v.Willisen	(73-76)
1977-1980	W.F.Berg	J.R.Günter	W.Balmer	(77-80)
1981-1986	J.R.Günter	R.Guggenheim	D.Gross	(81-82)
1987-	R.Guggenheim	R.Gotthardt	E.Mathieu	(83-)

TABLE 3:

Membership of the Swiss Society for Optics and Electron Microscopy

Year	Individual Members		Corporate Members
	Total	El.Microsc.	
1956	41		7
1957	61		11
1958	82		12
1959	109		16
1960	120		18
1961	135		21
1962	180		28
1963	186		32
1964			33
1965	201		38
1966			
1967	193		39
1968	194		40
1969	247		42
1970	303		42
1971	303	240	41
1972	318	255	40
1973	321	258	40
1974	350	265	47
1975	372	273	44
1976	428	312	45
1977	431	306	48
1978	441	328	55
1979	443	334	56
1980	465	354	61
1981	521	397	68
1982	538	415	72
1983	541	414	78
1984	551	419	79
1985	548	426	82
1986	560	435	80
1987	568	443	79
1988	617	473	84
1989	633	471	88

TABLE 4:

Meetings of the Swiss Society for Optics and Electron Microscopy

Optics	(jointly)	Electron Microscopy

Schweizerisches Komitee für Optik - SKO / CSO

	Optics	(jointly)	Electron Microscopy
1948:	Zürich, 23.4.48		
	Zürich, 4.11.48		
1949:	Berne, 25.6.49		
1951:	Zürich, 19.3.51		
1953:	Aarau, 10.12.53		
1954:	Zürich, 23.6.54	-----	Zürich, 23.6.54

Schweizerisches Komitee für Licht- und Elektronenoptik - SKO / CSO

	Optics	(jointly)	Electron Microscopy
1955:	Lausanne, 29.6.55	-----	Lausanne, 29.6.55
1956:	Balzers, 15.6.56		
1957:	Yverdon, 22.5.57		
1958:	Zürich, 18.11.58		
1959:	Neuchâtel, 26.5.59	------	Neuchâtel, 26.5.59
1960:	Zürich, 26.10.60		
1961:	Zürich, 8.6.61		
1962:	Fribourg, 9.11.62	------	Fribourg, 9.11.62
1963:	Berne, 19.9.63		Zürich, 22.-25.5.63 (jointly with DGE)
1964:	Berne, 15.10.64		
1965:	Berne, 20.10.65	------	Berne, 20.10.65
1966:	Basel, 4.11.66	------	Basel, 4.11.66
1967:	Zürich, 20.10.67	------	Zürich, 20.10.67
1968:	Berne, 4.10.68		

(continued on next page)

Table 4, continued

Schweizerische Gesellschaft für Optik und Elektronenmikroskopie
SGOEM / SSOME

1969:	Lausanne, 19.5.69	-------	Lausanne, 19.5.69
1970:	St.Gallen, 19.-22.5.70		(jointly with SFME)
	(jointly with DGaO)		
1971:	Fribourg, 8.10.71	-SNG-	Fribourg, 8.10.71
1972:	Dättwil, 22./23.6.72		
1973:	Lugano, 19./20.10.73	-SNG-	Lugano, 19./20.10.73
1974:	Heerbrugg, 30.10.74		Zürich, 30.10.74
1975:	Aarau, 3./4.10.75	-SNG-	Aarau, 3./4.10.75
1976:	Balzers, 27.10.76		Zürich, 14./15.10.76
1977:	Berne, 7.10.77	-SNG-	Berne, 7.10.77
1978:	Aarau, 11.10.78		Basel, 12./13.10.78
1979:	Lausanne, 5./6.10.79	-SNG-	Lausanne, 5./6.10.79
1980:	Berne, 15.10.80		Lausanne, 13./14.10.80
1981:	Davos, 25./26.9.81	-SNG-	Davos, 25./26.9.81
1982:	Luzern, 2.-5.6.82		Vaduz, 30.9.82
	(jointly with DGaO)		
1983:	Berne, 22./23.9.83	------	Berne, 22./23.9.83
1984:	Regensdorf, 14.9.84		Zürich, 5.10.84
1985:	Besançon (F), 29.5.85		Konstanz (D), 15.-21.9.85
	(jointly with SFO & DGaO)		(jointly with DGE & ÖGE)
1986:	Greifensee, 9.9.86		Berne, 10.10.86 (SNG)
1987:			Fribourg, 28.1.87
			(jointly with SAOG)
	Neuchâtel, 24.9.87	-SNG-	Neuchâtel, 24.9.87
1988:	Zürich, 5.10.88		Lausanne, 29./30.9.88
1989:	Neuchâtel, 21.9.89 (SATW)		Grenoble (F), 10.-12.7.89
			(jointly with SFME)
			Salzburg, 10.-16.9.89
			(jointly with DGE & ÖGE)
	Zürich, 27.10.89	------	Zürich, 27.10.89
1990:	Interlaken, 5.-9.6.90		Berne, 26.10.90
	(jointly with DGaO)		